THE LIVING HOUSE

George Ordish
THE LIVING HOUSE

Illustrations by
Alison Claire Darke

THE BODLEY HEAD
LONDON

To Roger Ordish

British Library Cataloguing
in Publication Data
Ordish, George
The Living House
I. Title
591.9422'3   QL 756
ISBN 0 370 30888 3

Printed in Great Britain for
The Bodley Head Ltd
30 Bedford Square, London, WCIB 3RP
by R J Acford, Chichester
First published in 1959
Second edition 1985

Contents

Illustrations

Preface

In writing this book about Barton's End, which is typical of many such houses in our countryside, I have been much helped by the published books and papers of many authors and I hereby tender my thanks to them. I have many people to thank for direct assistance: I am very grateful to those experts who read the typescript and made many useful suggestions, in particular to Mr Goodwin Bailey, Mrs J. Black, Mr G. E. Fussell, Dr A. M. Massee, Mr John Charlton, Mr E. Hyams, Dr H. L. Richardson, Mr N. Sloan and Dr C. Tanner. Finally I would like to thank my wife for an immense amount of valuable assistance.

GEORGE ORDISH
St Albans, May 1959

Preface to the second edition

This book is substantially the same as the first edition, with the story of the house brought up to date.

GEORGE ORDISH
Nancledra, Cornwall, February 1985

The People

John Barton   m. 1556   Mary Halstead
b.1537 d.1599   b.1537 d.1610

John	Jane	Elizabeth	Thomas Barton	Five other
b.1557	b.1558	b.1559	b.1560 d.1629	children
d.1558	d.1632	d.1559		

m.1584 Elizabeth Standish

Mark Barton m.1604   Margaret Stead   Six other
b.1585   b.1586   children
d.1631   d.1645

2 died in infancy   Elizabeth Barton m.1626   Robert Onway
                         b.1608 d.1688   b.1605 d.1660

Mark Onway m.1648 Prudence Sefton   2 daughters
b.1627
d.1665 of
the plague

James Onway b.1652, sold the property   Prudence
to a retired merchant 1689   b.1654

Joseph Munroyd m.1689   Ruth Breadman
b.1640 d.1705   b.1665

Isaac Munroyd m.1711   Anne Cobb   2 other children
b.1690 d.1752   b.1691

Charles Munroyd m. 1752   Susannah Aylesford,   2 other children
b.1716 d.1764   an heiress
                    b.1729 d.1800

Charles	George Munroyd m.1784	Sophia Holberg	Susannah
b.1752	b.1754 d.1808	b.1760 d.1809	
d.1752			

Charlotte Munroyd
b.1786 d.1817
Unmarried. Property sold by distant heirs to
a veteran of the Napoleonic wars, in 1818

Date	The House
1556–1599	The house completed in 1555.
1599–1629	Glazing completed. A porch was added to the house.
1629–1631	
1631–1660	A small cellar installed under the pantry. Kitchen made 1645.
1660–1665	Thatch renewed 1660.
1665–1705	Some bedrooms opened to the gallery.
1705–1752	Further improvements made; thatch replaced by tiles; ceilings to hall.
1752–1764	Further improvements; a new staircase 1752, ceilings in all rooms. Cellar enlarged, pump fitted.
1764–1808	Blocked fireplace. Coal burning stove.
1808–1817	

The People

John Hackshaw m.1797 Caroline Harris
b.1770 d.1833

William Hackshaw m.1819 Emily Hempstead   3 sons   2 daughters
b.1798 d.1852

Alfred Hackshaw   David Hackshaw m.1844 Victoria Kristen
b.1822 d.1898   b.1823
Disliked farming,
settled in Venice: Oscar Hackshaw
on death of his father   b.1845
let the farm to:

Emmanuel Burrows m.Marjory Rummer

and subsequently to another:

William Daker

In 1899 the farm was sold away from the house by Oscar Hackshaw, nephew and heir of Alfred Hackshaw. The house left unoccupied for 10 years.

Property bought very cheaply by an artist (1909):

Robert Dunchester m.1903 Myfanwy Evans
b.1870 d.1949   b.1884 d.1956

Robert F. Dunchester   Sloane Dunchester m.1935 Thomas Frazer
b.1910 d.1958   b.1912 d.1958   b.1914
m.1935 Louise Cavanagh (1)
      who m. (2) James Endicott
      no issue

Phillip Frazer m.1962 Desirée Barton   Caroline   Myfanwy II
b.1937   of Duxbury, Mass.   b.1939   b.1942
           b.1939   no issue   m.1967 Giles Thompson
           no issue

Phillip Thompson   Caroline   Sloane Myfanwy III
b.1969   b.1971   b.1974

Date	The House
1818–1833	Tank in roof in 1822.
1833–1852	Window in loft broken in 1833.
1853–1869	
1869–1899	Structure neglected.
1899–1909	During this period the house fell into very bad repair; the roof fell in at the north end.
1909–1949	Extensive repairs. Roof light to studio in north end. Central heating. Damaged by a flying bomb in 1944 and repaired.
1949–1958	Plans made to divide house in two in 1958.
1958–1984	House sold to Frazer Properties Ltd and made into an old people's home in 1965.

The reader must not expect in this work merely the private uninteresting history of a single person. He may expect whatever curious particulars can with any propriety be connected with it. Nor must the general disquisitions and the incidental narratives of the present work be ever considered as actually digressionary in their natures, and as merely useful in their notices. They are all united with the rest and form proper parts of the whole.

J. WHITAKER *History of Manchester*
(1771–1775)

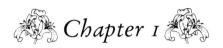

Chapter 1

A New Environment

LE CORBUSIER describes a house as 'a machine for living in'. This is indeed true and through the long history of man these machines have varied from the most simple—a few leafy branches leaning one against the other—to highly complicated ones, such as a Minoan palace or a modern skyscraper.

Man's houses, simple or complex, were built to provide him with shelter, to help the race survive and to enable him to live more easily. This last is the key to civilization, for when man did not have to spend every moment of his time in winning food or defending himself he had a surplus of time and energy to spare; many animals have this as well and fill the time with play. Man has done the same thing except that his play was more purposeful—at least some of it was—being devoted to improving his lot, particularly his houses.

Apart from being a machine, the house is also a new environment. The shelter provided by houses benefited mankind and thus provided for the extension of his species; but life is a powerful thing and constantly seeks to expand into any space where it can possibly be supported. Deep seas, high mountains, deserts, even deep polar ice can all support life of some kind, so that this new environment of the house soon had a wide variety of life in it, many forms taking advantage of the space, the materials, the amenities not being used by man. Some forms competed directly with him for these amenities—for instance the mice ate much the same food as he did and used wool and fur for nests, but none of the animals ate exactly the same things and required exactly the same accommodation as man. This was fortunate both for him and for them, for where two species of animals compete for the same materials a struggle ensues and one species becomes extinct. Had mice and men both needed exactly the same supplies one or other would have vanished, and it need not necessarily have been the mice.

It is quite easy for a species to become extinct; usually it is not some cataclysmic disaster or the attack of some virulent enemy that destroys a particular form of life, but the quiet and unspectacular extra efficiency of a rival species in obtaining the available food, water or living space.

All this can be seen to be happening in a house at any time, for during the lifetime of the place—from the creation of the new environment to the present—different kinds of life have both come into and gone out of it, as man changed his ways and as the ebb and flow of the evolutionary process worked over it. From the building of a house to its end, a wide diversity of creatures will have lived in it, and such changes are going on all the time as man adds new artifacts to his kingdom, or seeks to protect them and himself from the attacks of other forms of life.

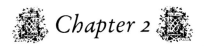

Chapter 2

Building and Populating
the Home

JOHN BARTON was the second son of Richard Barton, a farmer at Ashwell, a small village in the Weald of Kent, and he returned home in the spring of 1555, not having lived there for eight years, nor having seen his parents for five. He was eighteen years old and at the age of ten, rather late it was thought, he had gone to live on his uncle's estate near Guildford. In those days most children, other than those of peasants, were usually sent away from home to be brought up by strangers or relatives; it was a practice which was much condemned by visitors from abroad and was constantly cited by them as a proof of the callousness and indifference of the English. It was said that they sent their own children away and received in turn those of strangers, or distant relatives, because they could close their hearts more easily against a stranger's child than against their own and thus get more work out of these young people.

This may well have been a reason for the practice, even the principal one, though it was usually said that it was done for the purpose of educating the child and conditioning it to the stern responsibilities of gaining its living in a hard world. A residue of this custom may be seen today in the practice of sending the children of wealthier families to boarding schools. There is no doubt that it also had a considerable biological effect.

At that time the total population of England was very small, about four million in all. London was the only large city, with perhaps Norwich the next largest, and in the rest of the country there were only small towns and villages, which latter lived a life of almost complete self-sufficiency. The inhabitants knew they must grow their own food: the clothes they wanted were carded, spun, woven and made from their own flax and wool. They made their own beer, wine and verjuice. If the harvest failed, they went hungry and if it was abundant small surpluses, wool, hides, or cloth, could be dragged out of the villages and traded in the market town for outside goods—first for necessities such as salt and metals and then for such luxuries as spices, oranges, dried figs and so forth. Many peasants never left their villages even to go as far as the next one; consequently the class of peasants and small copyholders was very inbred. Anyone in this village stratum of society was perforce related to everyone else, and most of the population did live in villages. The Bartons were not exactly squires but successful farmers and on their way to the squirearchy. Kentish farmers regarded themselves as a cut above the average run of farmers in the country and, in fact, they still do.

Among the upper classes, the nobility and gentry, which by the end of the sixteenth century consisted of some sixty-one titled and noble families, five hundred knights and sixteen thousand esquires, the custom was not only to send one's children—both boys and girls—away from home at about the age of seven but also to arrange their marriages for them. This tended to produce on the one hand a somewhat slow inbred peasantry and on the other a more mixed and vigorous extrovert squirearchy and nobility. These biological effects may well have produced the ready acceptance of the state of servitude by the peasant which was a characteristic of those times, because though there were peasant revolts during this period they were comparatively rare occurrences.

Richard Barton was a wealthy man, firmly attached to the Protestant cause, and could not only read and write himself but had a wife who could do likewise, and who had seen that his seven surviving children (three sons and four daughters) out of twelve could do so as well. 1555 was not a prosperous time, particularly for Protestants and in Kent, because Wyatt's rebellion, which had started in North Kent the previous year, had failed to prevent the marriage of Queen Mary to Philip of Spain, and the persecution of Protestants was now in full swing. The Bartons were in no way implicated in the rebellion, but being of that faith, they found it wisest to remain quiet, particularly as the name Barton somewhat suggested troublemaking after the incident of the Maid of Kent in Henry VIII's time. This was when a maidservant at Allington, Elizabeth Barton, had had fits and seen visions; becoming a nun at St Sepulchre, Canterbury, she made a large number of prophecies and acquired an immense influence over events. She opposed the dissolution of the king's marriage with Catherine of Aragon; she persuaded Archbishop Warren to withdraw his consent to conducting the marriage service for Henry and Anne Boleyn, and she was active with all those opposing the separation from Rome. Eventually she was hanged as an impostor with some six others (all men) at Tyburn in 1534. One can see that the Protestant Bartons had cause to be wary in 1555, even though it was twenty years later and Elizabeth Barton had supported the Roman Church. So Richard Barton concentrated his energies on improving his farm and let the great world roll by unheeded.

It was in fact the foundation of his family's prosperity. Barton and his children indulged in a number of activities which were thought strange and very advanced by their neighbours; for instance they read textbooks on agriculture, such as Fitzherbert's *Boke of Husbandrye*, a work which became very popular and ran to eight editions before the end of the century. Barton prospered because he had grasped the importance of the recent advances in the art of agriculture—it would be called scientific agriculture today—of which the most important were enclosure, sheep farming and good relations with his labourers and copyholders. With his eldest son, also called Richard, he had enclosed a large area of the ancient 'waste', felled and cleared part of the forest and intended to settle the young man on his new farm, Bartons, after his own death.

Two good wool crops, sold to the merchants at Rye for export to Flanders, now gave him the opportunity to do the same for his second son John. It was for this reason that he had called him home and told him he should marry Mary Halstead, a young girl of sixteen from a good family in the neighbourhood, who had been in his house for the last eight years. She would bring a jointure of £100, a considerable sum in those days, which would enable them to build a house and make a new farm to the south of Bartons to be called Bartons End. Old Richard Barton hoped that the young people would agree to the proposal and Mary hoped the young man would like her and love her. She may also have hoped that she would be able to like, even to love him, but under the circumstances she would have been unlikely to oppose the suggestion. Marriage was the only career a girl could follow and the possibilities of a new house and a new farm of one's own were very attractive.

For some months the family had been busy selecting the site for the farmhouse: it had to be convenient for the fields, and these fields in their turn had to be selected so as to cause the minimum of disturbance to the commoners. In the long run one could enclose as much land as one wanted by one means or another; for instance, one could buy up the strips, or parts of the strips, in the common fields belonging to the local village, or one could threaten or cajole the owners, but Barton was a great believer in doing things as quietly as possible and he was able to make John's farm mostly from the waste and from the forest. The farmhouse had to be sheltered, on a well-drained piece of land, so that it could be approached in winter and, above all, it had to have water. A good well was essential; consequently after three sites had been selected as possibilities, a water-diviner was asked to judge them for water; and before the amount of timber needed for the house was even calculated or selected, the well was dug on the site the diviner selected. It proved to have a good and abundant supply and caused much rejoicing, for 'fair water' was a constant difficulty of the early housewife.

Mary Halstead naturally took a great interest in all the preparations for the new house. She did not know if John would accept his father's proposal and, though all the Barton family acted on the assumption that she was to be the mistress of the new place, she herself could not help wondering if it were wise. She made suggestions as to the design and placing of the house, but she was too young and too uncertain of her position to insist on too many modern innovations. Her proposal that the building should face the south with a view over the valley was laughed to scorn: that would let in all the sunlight which would fade the hangings and clothes, warm the house and turn all the food sour in the summer. A house should run north-east and south-west so that the early sun would light the bedrooms and get people up early, and the setting sun should light the living rooms and kitchens so that work could be continued as long as possible. If a house had to run east and west it should certainly have a blank windowless wall on the south side in order to exclude the sun.

Bartons End

The site had been selected and the well dug: the next thing was roughly to sketch out the house and to mark it out on the ground; once this had been done the quantity of material needed for the house could be calculated. The main items were wood for beams, rafters and boards, bricks for the fireplaces and chimneys, stone flags for some of the floors and thatch for the roof. The biggest item of all was the wood; although there was plenty of wood available, the difficulty was in sawing it, for it was the custom to use oak almost exclusively, and this is an exceptionally hard and difficult wood to handle. The estate belonging to the local squire, a distant relative, had its own sawpit where skilled sawyers laboured hard and continuously, and every day the big cross-cut saw had to be sharpened: the sawyers developed enormous muscles and had to be particularly well fed to support such arduous work. 'Strip whilst you're cold, and live to grow old' was the advice given to the young man coming fresh to the task. The squire agreed to help, and the preparation of the beams and boards went on for the whole of the winter of 1554–55. The building of the house actually started in the spring of 1555. The matter was greatly helped by Richard Barton's being able to load a wool waggon returning empty from Rye with a supply of old ship's timbers: these were always much in demand for house building and they frequently supplied the first life in the house, life which had begun even before the house was completed. It began indeed as soon as the first beam was in position, whether it was a comparatively new one, or one of the old ship's timbers.

This life was made up of the various wood-borers in the beams as they were put into position and then the spiders that spun webs in the house as it was being built and the insects which were caught in the webs. As the house rose slowly upwards, so more and more different forms of life were attracted to it, and with each animal that came in there also came numbers of parasites and predators which only lived because of that animal. A series of passing insects went through the house as it was being built, wasps, bees, butterflies and flies, according to the time of year and the weather. The first vertebrates were frogs clustering round the well-head and in the cellar. The first warm-blooded animals were house-sparrows attracted by the crumbs from the builders' midday meals. The first worms were parasites of the sparrows. The flag showing that the ridgeboard was in position was not hoisted till August 1555, and as soon as the roof was finished the windows were put in and the house sealed up till the following spring. Thus it was the first mammals to live in it were John Barton and Mary Halstead, who had renewed their childhood friendship of eight years before, agreed to Richard Barton's proposal and were married in March 1556, with much feasting, rejoicing and the coarse but kindly bucolic humour of the age.

On their new farm they lived a happy and complete life; it was an extremely hard one, but unremitting toil was what they expected. While, on the whole, they dominated all the life in the house, there was much of which they were not aware and a great deal of which they were only vaguely conscious. With some of it, such as rats and mice, they were constantly at war and to others, such as butterflies and spiders, they were indifferent. Life is a cycle of birth, growth, reproduction and death. The pattern of living in the house may be said to be a series of circles, slowly moving, expanding, changing, declining and merging with that of the human beings, the dominant factors, so that the history of the house is a figure resulting from the inter-movement of the various circles of the life forms in the house.

As time went on, over more than four hundred years from the house's construction to the present day, some new circles were brought into the pattern, such as those of the cockroach and bat, and others, such as the bed-bug, were taken out of it completely or, as in the case of wood-borers, almost completely. In the following chapters I give some account of these circles and how they waxed and waned.

It is characteristic of all forms of life that they have the urge to reproduce themselves and to the greatest possible extent. The lower the form of life the more offspring are produced, so that if there were no checks on the survival of a species it would rapidly cover the entire globe: the highest forms of animal life are the slowest breeders. Charles Darwin gives the example of the elephant:

> The elephant is reckoned the slowest breeder of all known animals, and I have taken some pains to estimate its probable minimum rate of natural increase; it will be safest to assume that it begins breeding when thirty years old, and goes on breeding till ninety years old, bringing forth six young in the interval, and surviving till one hundred years old: if this be so, after a period of from 740 to 750

years there would be nearly nineteen million elephants alive, descended from the first pair.

If the population of such slow breeders as elephants can grow at this rate, think what the population of mice would become when a female will have about six litters a year, were there no checks to their expansion.

The fact that more progeny are produced than can survive led Darwin to propound his theory of natural selection and the survival of the fittest. It is obvious that as food is scarce those progeny which are best adapted to obtaining it are most likely to get it and thus to survive, and however heterozygous an animal may be, some of its offspring will resemble it. Consequently any favourable combination of genes will tend to be passed on to all subsequent generations. This also applies to all favourable mutations: these again will tend to appear in the offspring and if advantageous to the race, will persist in it. There is constant competition among all animals, both within a species and between species, for the scarce resources of the environment and that is why few animals have time for any activity—such as play—other than food-gathering and reproduction.

The same immutable laws may be said to have applied to the Bartons as well. They survived and prospered, firstly, because man was a successful animal and dominated the other animals of the environment, and, secondly, they survived in the competition among their own species, among men, because they were adaptable and could use their intelligence to take advantage of the conditions they found present.

Their chief weapon was enclosure. Enclosed land would produce bigger crops: one could keep better cattle on enclosed fields than on the commons, and surplus crops and animals were wealth which could be exchanged for goods or even intangibles, such as raising oneself in the social system by acquiring a knighthood, or by marrying a child into a noble family.

In the animal world the checks on the indiscriminate reproduction of the species are the lack of food, oxygen, and a suitable environment, and also the attacks of enemies such as parasites, predators and diseases, and it is these factors that controlled the volume of life in John Barton's house through the four hundred years of its existence.

The house was built to provide a suitable environment for John and Mary Barton; to it they brought their food and they obtained their oxygen as a free gift from the atmosphere outside. They and their human successors differed from almost all the other animals in the house, in that Mary and John's sole object in life was not the production of offspring, but also the worship of God, the embellishment of life by the arts, by enjoyment and social intercourse. Mary gave birth to a child every year, which was about as near to maximum production as possible, but this annual event, which was a commonplace for that age, cannot be considered as the only object of their existence, as it can be of the brute creation.

The farm produced wheat, oats and barley, and meat in the form of oxen and sheep. The wool crop and a little wheat and cheese produced the Bartons' cash income which was needed to buy iron from the Weald smelters and specialized tools such as the saws, bits and augers to work the brick-like oak of their forests. They only knew milk in the spring and summer flushes and they made much of it into cheese. A minor, though important by-product was the earth from the stables, because this absorbed the urine from the animals and the salt, potassium nitrate, formed in it. This earth was in great demand, not as a fertilizer for the fields, as might be imagined, but as a source of saltpetre for the manufacture of gunpowder. There would not have been enough of it to make much difference to the fields but such earths were the only way to obtain this essential ingredient so needed by the armed forces. In fact the Queen's nitre men could commandeer any stable earth anywhere, dig it up and wash out the nitre and no one could resist them or complain of interference. Naturally, they rarely visited the farms but obtained most of their supplies in the town, where there were big concentrations of horses.

The food needed by all the animals in the house was of four main types—carbohydrate, protein, the mineral supplements and vitamins and, of course, water, which though hardly a food, is yet the medium by which the cell works. The most important and the most difficult to obtain was the protein. Proteins are a class of organic chemicals whose essential feature is a content of some fifteen per cent of that almost inert gas, nitrogen, and it is again a curious matter that this very shy chemical, nitrogen, should be so essential to life. The very life substance itself, the cell nucleus can divide and cause the animal to grow; hence nitrogen is one of the factors limiting the expansion of animal populations. Although the atmosphere in which animals live is four-fifths nitrogen, they are unable to use it in this form (except a group of bacteria found in association with leguminous plants) but must obtain it already combined with other substances—usually as protein in plants or other animals. Nitrogen, however, can be made to combine with other gases, such as oxygen, if submitted to great pressure or if passed through an electric spark. This happens in a flash of lightning and a modicum of combined nitrogen falls with the rain in a thunderstorm and enriches the earth. This dull, lifeless chemical—nitrogen—which is so essential to life is possibly so because of its very dullness. When it combines with other chemicals it is only held in a loose bond by them, so that, for instance, a substance like potassium nitrate, a chemical containing potassium, nitrogen and oxygen (KNO_3) can be absorbed by the roots of a plant and easily broken up by the chlorophyll and sunlight in the leaf because the nitrogen is held so loosely by the atomic attractions within the potassium nitrate molecule. It is for the same reason that nitrates are used in explosives; the molecule is easily broken up and releases oxygen to feed the explosion. Moreover the nitrates are readily soluble in water which enables them easily to move into the plant.

'All flesh is grass' and no animals are able to live in the last analysis except by eating plants: even a carnivore such as a lion lives by eating animals which live by eating plants. It is only plants which are able to take simple salts in the soil and combine them with oxygen from the air and form protein and carbohydrate. They do this by means of the action of sunlight on the chlorophyll in the leaf; a process known as photosynthesis.

Even the plants cannot use the nitrogen of the air but must obtain it in the combined form from the soil and the soil has three ways of obtaining it, namely (i) from rain water, (ii) from the decay of organic matter in the soil, such as faeces, plant and animal remains, and (iii) by the action of certain bacteria (Pseudomonads) which live in the nodules of the leguminous plants and do have the power of directly using the atmospheric nitrogen.

Living things are not only characterized by living but also by ceasing to live—by dying—usually by falling a victim to some other form of life, for this is the case whether they are killed young as food for another animal, or become enfeebled by old age and fall a victim to some disease. It is of course possible for an animal's organs to be so weakened that it just fades away, but usually there is some life form in the neighbourhood which, taking advantage of this enfeebled condition, profits itself by feeding on and killing the weakened individual. The reverse process then starts and the complex chemical compounds of the body break down again to the simple elements from which they arose, though here again there may be a chain of life-forms involved. Of course the living body is also taking in complex compounds as food, using some and breaking the rest down into simpler ones all the time, the simpler products being disposed of in the expelled breath and excrements of different kinds.

In the pattern of life the process of decay is indeed as important as the synthesis of the compounds itself—a fact recognized by the sculpture of Germaine Richier, which usually suggests the breaking up of her subjects into their elementals.

Life is a flow of certain chemical atoms to a constantly rearranged pattern. Some millions of tons of nitrogen, carbon, oxygen, hydrogen, phosphorus, potash (and other elements) are constantly being assembled by the life process and made into complex compounds and then being broken down again, often to the basic elements themselves.

The Buddhists and Theosophists believe that the soul is reincarnate time after time in animals and men. It is difficult to know about our souls, but our bodies are constantly using elements that have been used before. Ants after feeding on sugar or nectar regurgitate a drop, hold it in their mouths and pass it to another ant. If you feed a few ants with a blue-dyed sugar solution, soon all the ants in the nest will be blue. This seems a rather revolting practice, but it is what we all do with our air. At the cinema, the cocktail party, board room or lecture, we are constantly re-breathing other people's (so to speak) regurgitated air; moreover the process stretches backwards in time. As Julius Caesar fell dead to the floor of Pompey's theatre, he cried with his last breath, 'Et tu, Brute!' and he expelled a few cubic centimetres of oxygen in doing this. Those same cubic centimetres of oxygen have been moving around in the world ever since, and one is quite likely to use some atoms of them oneself in the course of one's lifetime; for a man living to seventy will breathe some 800 tons of oxygen in the process.

DOWNSTAIRS PLAN

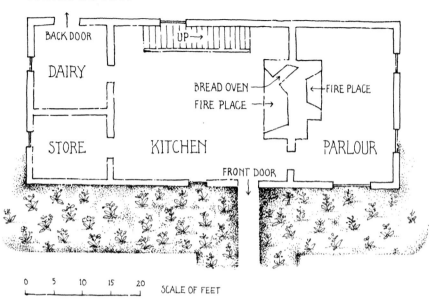

BACK DOOR

DAIRY

UP →

BREAD OVEN ←
FIRE PLACE → ← FIRE PLACE

STORE   KITCHEN   PARLOUR

FRONT DOOR

0   5   10   15   20   SCALE OF FEET

UPSTAIRS PLAN

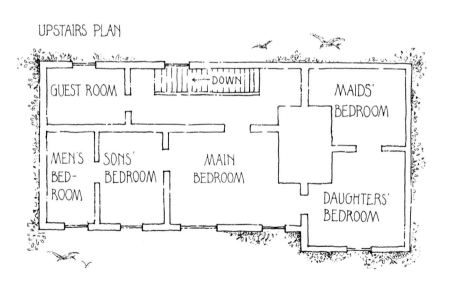

GUEST ROOM   ← DOWN   MAIDS'
                                   BEDROOM

MEN'S   SONS'   MAIN
BED-   BEDROOM   BEDROOM
ROOM
                                   DAUGHTERS'
                                   BEDROOM

Bartons End. Ground plan

The Bartons were a progressive family who were prepared to accept a number of new ideas when building John's new house. The yeoman's house in those days usually consisted of a large hall or kitchen with a huge brick fireplace at one end; behind this was a parlour for the master with a second fireplace the other side of the stack. The hall, being open to the roof, was a survival of the days when a fire was built in the centre of the floor and smoke allowed to escape from a hole in the roof. At the other end of the hall there might be another room built out to form a pantry, or there could be a lean-to at the back for the same purpose. Built over the parlour would be two or three bedrooms opening one from the other, to which one obtained access by a ladder or steep stairway. In the case of the parlour, the walls of the ground floor were built in the usual way of timber-framing which reached to the floor of the first storey only; on this a main beam or so would be laid and on these the floor joists, which projected through the wall to just over the lower storey, were stretched flat. On the jutting-out portions, known as jetties, the upper wall was built and the weight of the walls served to counterbalance the weight of the floor, people and furniture within the house on the first floor. The floor of the upper rooms served as the ceiling for the room below and as these were usually of oak, the rooms tended to be very dark.

The Bartons introduced a number of innovations to the accepted plan, for they provided a bedroom for the male servants (who previously used to sleep on the floor of the kitchen), a guest room, in the hope that one day John would have to entertain an important man, and a staircase leading to a neat but useless gallery running to the guest room. The gallery was to become of immense value to subsequent occupiers of the house as it enabled them to modernize it with very little effort.

As will be seen from the plan of Bartons End (page 26), the bedrooms all opened out of each other. The principal bedroom led to two others, one beyond the other, on one side, and to two more on the opposite side, which was done partly from precedent and partly to give the master control of the life of the house. The pair on one side was for the sons and the male servants; the pair on the other side was for the daughters and the maids. When the time to go to bed arrived, the master of the house gave the signal, whereupon first the maids went upstairs to their room, carrying a rushlight to supply illumination; they were followed by the daughters, the menservants, the sons and finally the master and mistress. In the early morning, about dawn, the mistress woke her daughters and maids whilst the master aroused his sons and menservants, for all had to be at the day's hard toil early and long.

The neighbours thought the Bartons were being very pretentious with their gallery and, above all, by providing a room for the sons and for two old farm-labourers, but Richard Barton was able to tell a hawk from a hand-saw; he saw that John would not only please his men but also have control over them more readily if he knew what time they went to bed and got up. Moreover, the experience of these two old labourers would not come amiss in directing his son's new enterprise.

Another innovation was in the method of building the house, which had been suggested to Richard Barton by a friend of his, a surveyor, as the architects of that day were called. Having accepted the idea of the main hall of the building being made into two storeys, rather like a town house, the innovation consisted in carrying up the main uprights of the building from floor to eaves, thus doing away with the overhanging outer jetties. The floor joists were then mortised into the horizontal wall beams and the whole house considerably strengthened. There were a great many consultations before the carpenters had grasped this fundamental change of idea, but once they had they became enthusiastic about it, as it meant they could cut the floor space of the house about as they wished, for the floor now no longer had to be whole to balance the weight of the upper walls on the jetties. They soon saw they could have a better staircase and the gallery to the guest room for John's hypothetical important visitor.

In the forest during the winter of 1554–55 the great trees were felled, to be dragged down the muddy lanes by the slow but powerful oxen to the sawpit at Fordover. All that winter and well into the spring the sawyers were hard at work—the labour was immense, but it was a good way of keeping warm. It was not the custom in those days to spend a great deal of time in seasoning the timber and very little warp or twisting took place after it was in position, perhaps because of the very solid construction used or the fact that any gaps or twists would be made good by fitting in, or otherwise amending the structure.

The carpenters took over the beams and scantlings from the sawyers in the spring of 1555, and by the side of the sawpits they started to make the framework of the house. The beams were trimmed, the mortises and tenons made and the whole framework of each side was assembled flat on the ground of the near-by meadow. All the joints were numbered with Roman numerals as not only was this customary but very few artisans at that time were able to read the newly-introduced Arabic figures. Moreover, the Roman numbers were readily cut with a chisel. Many of these numbers can still be seen on the beams at Bartons End today. The mortises were held by wooden pegs, just as they are today, though such big timbers are seldom used now, and the holes were offset so that the joint tightened as the peg was driven, again in the same way as our smaller joints are tightened nowadays. Whenever the weather was fine the carpenters carried their beams out to the field and assembled them there, and the family watched the growing house with quiet excitement. In the meantime the bricklayers were constructing the giant chimney on the site, around which the house would be built. When all was ready the pegs were knocked out of the joints and the whole carried to the site. Each side was again reassembled flat on the ground, when the numbers cut on the beams were invaluable, and the complete side was then hoisted up into the air with much labour and groaning. Each side was temporarily supported and soon the whole framework was in position, then the rafters and ridgeboard were completed and the chief carpenter, Harold Leigh, nailed the flag to the roof on August 12. Richard Barton gave a big party and announced the forthcoming marriage of John and Mary.

The thatching of the roof was completed by November 1555, and the glaziers had completed the windows by the same date; they contained only a few leaded lights in the top; the rest was closed by a wooden shutter, or by a frame covered with oiled linen.

By 1 December 1555, the house may be said to have been completed and we must consider its first inhabitants. The claim to this distinction belongs to the wood-boring beetles, because they were actually present in some of the timber when it was used in building the house.

Chapter 3

The First Inhabitants—
the Wood-borers

IT IS a commonplace of entomology that the life of insects is so different from man's that although it is easy enough to describe, it is almost impossible for man to envisage and appreciate. It is a form of life in which the animal is quite unaware of the reasons for what is happening around it, of what has gone before, even in its own life-cycle, or what will occur later. At the same time insects may be aware of things of which we are not. For instance, bees can sense the plane of polarized light, some insects see things by means of ultra-violet light which are invisible to us, other insects can hear sounds beyond the range of the human ear, and most insects have a far more sensitive sense of smell than man. Moreover, many have the power of smelling by touch—the chemotactic sense—which assists them in finding food, mates or suitable environments. It is a life in which the response to any given stimulus is almost as automatic as pressing the starting button of a motor-car and where things are done to a pattern which, for want of a better name, we call instinct. It is a world where one might say there is little learning and much mechanics: that is, a mechanical response to the environment, not a reasoned one. It is a life with much waste in it, for few individuals survive from the thousands of eggs produced, in striking contrast to the many survivors of the few young produced by mammals, particularly man, other primates and bats.

Although we can see that there is much waste, this is only from the point of view of any one particular species; from the contrasting point of view of life itself there is scarcely any loss, for the waste forms a basis for the nourishment of other forms of life. Insects feed on each other; spiders feed on insects; birds feed on both, cats feed on birds; men eat birds and so on. Elaborate food chains are established all starting from the ability of the plant to make food and leading ultimately to the nourishment of the 'higher animals', particularly man, who in the last analysis lives on the wastage of other forms of animal and vegetable life. Well may Genesis say men '. . . have dominion over the fish of the sea, and over the fowl of the air, and over the cattle, and over all the earth, and over every creeping thing that creepeth upon the earth.'

In the very different world of insects, sometimes the response does not take place, as sometimes the engine may not start when we press the button. When confronted with an unusual situation—in fact a problem—an insect may well have no instinctive pattern to guide it. As a whole, insects are considerably inferior to mammals in solving problems, though most insects are able to learn to some extent and to profit from their learning; for instance bees and wasps can find their way back to their nests over long distances by learning the way after successive explorations. Anything an insect does is done because it will be beneficial to the future of its species—it will add to the chances of survival of itself or its offspring: in general insects are so prolific and the struggle for existence is so intense that there can be no time for any activity which does not bear on survival, because those kinds of insects (species or mutations of species) which do indulge in such inessential activities are most likely to become extinct and have their place taken by others more survival-patterned. This is not a conscious activity on the part of the insects; it is merely the current conditions imposing a pattern of behaviour on a species or extinguishing it.

With insects, out of the many thousands of young produced only a few will survive to continue the race; this leads to the rapid adoption of characteristics favourable to the survival of the species, which then become instinctive and lead to the creation of new species when the new characteristics become so different as to prevent breeding between the new and the old.

There are a few exceptions, when an insect's activities are not directly beneficial to the survival of the species, such as the autumn playtime of the worker wasps, in which they stop collecting nectar for the nest and indulge in an orgy of eating the food they collect, just before they die of cold or exposure. Such playtime is rare in the lower animals: in the case of the worker wasps it has no effect on the race because it takes place after the virgin queens have flown, which is perhaps why it can be permitted.

The great antiquity of insects, their fecundity and their quick life-cycles compared to that of man mean that a pattern of life is rapidly fixed on a species, which will enable it to exploit existing conditions and if these conditions alter, then to change to meet new circumstances. If any particular insects cannot do this they disappear as a species, which is how the instinctive pattern arises.

It seems very wonderful that the hunting wasp will dig a burrow, find a caterpillar, paralyse it with a sting, drag it home and lay its eggs on the paralysed host, so that the wasp grubs will have food when they hatch, but if it did not have this pattern of behaviour it would not exist as a species. That the female wasp can do all this for grubs it will never see and of which it is never conscious is indeed strange, but no stranger than that the human embryo grows into a human and the acorn embryo into an oak.

In spite of an apparent complexity in passing through a number of different stages and environments, such as egg, larva, pupa and adult, perhaps crawling, then flying, an insect's life is a comparatively simple one. This is because nearly all its behaviour is governed by the instinctive pattern set on it by generations of precedent—one might say that it is a real conservative. Nearly all the situations which arise can be dealt with automatically, for 'problems' are comparatively rare and insects are not individually very clever at solving them: but in the main they may solve a problem because there are so many of them and out of millions some will have hit on the right answer by chance, like the eight monkeys who would—given time—eventually write all the books in the British Museum.

To take an example—the codling moth emerges in the early summer, mates, and the female seeks an apple on which to lay eggs: she is guided by smell and, if she cannot find an apple, has a problem to solve. Some few moths, faced with this situation, tried pears, where some of the eggs hatched, larvae were reared and the pear soon became a host for the codling moth. The species, through solving a problem, thus extended its territory which it is now attempting to enlarge yet further, for apricots and peaches have recently been attacked by this insect. It also lays eggs on leaves and shoots of apples and has succeeded in penetrating these latter, though not as yet in completing its full life-cycle from this new environment. It has solved a problem, though not in the same way as a mammal would have solved a similar problem. Not all problems can be solved, for, faced with the experience of a sheet of solid air, a fly does not know what to do, but continues to buzz up and down a window till exhausted, whereas men have escaped from the most extraordinary predicaments because their instincts did not bind them to a pattern of behaviour. However, it is only a question of degree, because new circumstances frequently arise where man does not know what to do. For instance, he has found no way so far of dealing with the threat of nuclear warfare.

Some of these points can be appreciated in describing the wood-borers that went into Bartons End.

The trees were felled in the forest which had been cleared to make part of the farm. The trimmed trunks lay from a few weeks to a year before they were hauled to the sawpit and this gave the first wave of insects an opportunity to install themselves. When the trees were cut they contained a few tiny bluish-stained tunnels belonging to the pinhole borer beetles. The fertilized females of these insects flew on to hardwood trees and started to bore a tunnel, frequently with the help of the male, into the sapwood of the trees. After a considerable penetration they laid eggs in the tunnel and they also introduced a special fungus around the eggs. As the eggs hatched the fungus grew and the young larvae fed on it and thus returned along the route opened for them by their parents, although they had never seen them. The young grubs, who cannot live on wood alone, co-operate with the fungus so that the wood is partially digested for them, but the fungus will not grow unless the moisture content of the timber is high, and when the trees are felled they soon begin to dry out. A characteristic of the pinhole borer attack is that the galleries are straight, stained and free of the dust that is usually found after other woodworm attack. After the trees were cut, the fungus soon stopped growing and the larvae started to die so that there were only very few live pinhole beetles in the house by the time John and Mary moved in, and there were no live ones left after they had been there a year.

However, the timber was not then free of borers because as it lay in the forest it was attacked by the powder post beetles. At the time Bartons End was built there was only one species of these *Lyctus* beetles in England, though now there are six. While the timber lay in the forest and the saw-mill yard, the adult fertilized females of the dark *Lyctus* were flying around looking for suitable wood unprotected by bark in which to lay their eggs, testing the wood by tasting it. They were very particular in that they would only lay eggs on wide-pored hardwoods and only in the sapwood of such timbers. In the almost untouched forest the beetles did not thrive very well because they only had old fallen trees or branches in which to lay and these were few and far between, but in the Bartons' wood-yard the sawn timber lay about awaiting use, which gave the females far better opportunities for finding the right places for egg-laying, and the sapwood of the Kentish oak was ideal for this purpose.

The long, tapering, cylindrical egg was pushed by the female into a sapwood vessel because the larvae will only thrive in sound timber having a high starch content, and the sapwood is the most recently formed wood of the tree, which is still living and contains this vital food. Each beetle can lay up to about fifty eggs but not all of these necessarily hatch out to young larvae. Some of the eggs failed to hatch, some fell victim to other insects, some were dried up by the sun and others were drowned by the rain.

When the egg hatches it is already in a wood pore of the timber and the young grub, as it comes from the egg, finds itself in a cell much bigger than itself, which means it cannot start its life of boring into the wood and feeding on its favourite food, the starch (though it must obtain some protein as well), because it is too small to get a grip on the walls of the cell pores and thus attack the wall. The larva first feeds on the remains of the yolk of the egg and such starch as it can find in the wood pore, and then when it is the same size as the pore it starts boring down into the timber. As it forms its gallery it leaves behind a fine powdery dust, and as it forges on, always in the sapwood, it grows in size. About a year is needed to complete the life-cycle, so that by May the fully grown larva is approaching the surface of the timber again, when it forms a cell and pupates, emerging in June as a fully grown *Lyctus* beetle. The beetles take to the wing, mate, and the females set about finding another piece of wood with wide pores and a suitable starch content in which to start the cycle over again.

Lyctus is essentially an insect of unseasoned sapwood and only very little of this went into the house, so that not many of their species were present. Some of the flooring in the passages was made of this material and had *Lyctus* larvae in it as it was laid. However, their numbers soon died down, because as the sapwood is living, the tissue transpires and uses up the starch in the process, which meant that after about two years there was not enough starch left to make the wood sufficiently attractive for the egg-laying female beetles as they appeared in the summer months. These adults either escaped from the house and sought the forest or laid eggs on the log pile and kindling wood and were burnt in due course. After four years, in 1560, there were no more live *Lyctus* in the structure of the house, though they continued to be brought in with the fuel.

Some of the timber used in the house was second-hand, as has been mentioned before, having been obtained by Richard Barton from Rye where it had once been part of a ship. It was perfectly sound as it had not been external timber and thus subject to the attack of the marine borer, *Teredo*, but had come from the captain's cabin. The Squire had carefully examined it before purchase, as he was well aware that wormy timber was weak and undesirable, though he had but little idea whence the worms came; this timber, however, did have one defect—it was about sixty years old and just at the right stage for attack by the furniture beetle (*Anobium punctatum*). These beetles lived in the same woods from which the house timbers were cut, where they attacked fallen trees and dead branches, and they led a precarious existence, again because of their somewhat specialized feeding requirements. They would only penetrate softwood some twenty years old, and hardwoods of sixty years or more, because the timber did not reach the right condition for the insect until after this period.

Once the right timber had been found by the adults this particular race of beetles was able to thrive, as the larvae had adapted themselves to eating cellulose; they thus had an advantage over other timber borers, such as *Lyctus*, who could not digest this difficult material.

The furniture beetle, or *Anobium*, is a small light-brown insect which varies in length from one-tenth to one-quarter of an inch. The freshly emerged insect is covered with a light, yellow down which becomes worn as it ages. The wing cases have a number of rows of pin spots on them (which give rise to its Latin specific name of *punctatum*; the generic name—*Anobium*—arises from its habit of shamming dead) running the length of the body. The head is curious in being shielded by a hard hood or lump on the insect's thorax, the middle segment of its body, which partly protects and conceals the head.

The furniture beetle adults emerged from the pupal cases in June and lay quiet in their burrow for a few weeks in order that the newly formed parts might harden and take on their brown colour. Some were males and some females and when they were mature they ate their way out of the pupal chamber, leaving behind them the characteristic round exit hole about one-sixteenth of an inch in diameter. Some of these new beetles mated almost at once, frequently with the female hiding in an old emergence hole and the male making contact with her from outside, and some of them flew off before mating; the primary duty and desire of the females was to find suitable sites for their eggs. Most of the insects used the same trees from which they had emerged, but once the beetles had taken to the air it was difficult for them to find the site again and some roamed quite far on the breeze. Most of these failed to find a suitable site and just died from exhaustion or were eaten by birds or other animals, but just beyond the edge of the wood was the meadow where John Barton's house was being built, a great expanse of bare timber was on view and could be imagined as a very paradise for *Anobium*, were the insect capable of such a concept.

Some of the far-flying females settled on the rising house, but did not find the wood very suitable, as most of it was too green to bring about the proper egg-laying reaction in *Anobium*. However, those that found the old timber brought up from Rye were lucky and started to lay their eggs in the cracks on the rough surface of the wood. These were deposited in groups of three or four; they were a dirty white in colour and the shape of a lemon, and about a third of the surface at one end was pitted with little marks while the rest was quite smooth. As the female

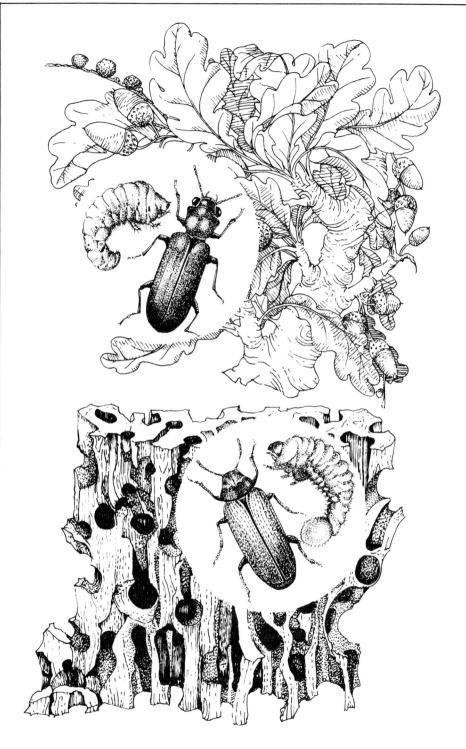

*Wood-borers; the powder post beetle, attacking green oak timbers,
and the furniture beetle*

pushed the eggs into the cracks they were often squeezed out of shape, but this did them no harm. The number of eggs laid by each female depended on the amount of food reserve with which they had started life and the amount of hunting they had had to do before they found suitable sites; between ten and forty eggs were laid by each of perhaps a dozen insects, so that the amount of infestation going into John's house at this moment was not great.

At the end of two weeks the eggs had started to hatch from the bottom and the young grubs immediately bored into the wood. Their mothers had selected the sites most efficiently, for it would be very unusual for an *Anobium* to lay eggs where the grubs could not at once obtain food and shelter. The small naked unprotected grub is at its most vulnerable state at the moment of hatching: it could be attacked by hunting insects, or it could be dried up by exposure to sun or wind. In the struggle for survival over the thousands of generations of *Anobium*, those females which were careful to lay their eggs near suitable food and those that laid eggs that hatched from the bottom were more likely to have offspring that survived. These offspring in turn were likely to have more descendants with these two characteristics until in time this behaviour became the pattern for the species—it had become instinctive for them to act in this manner.

The larvae were greyish-white in colour with brown jaws and were covered with fine hairs. They were so small that the hole they made in entering the wood could not be seen with the naked eye—the holes we do see are made by the adult insect coming out. Once safely inside the wood they commenced their long, uneventful existence of eating and growing in perpetual darkness, getting all they needed from the wood itself—air, water, carbohydrates, protein and vitamins. The hairs help the young larva to get a grip on the wall of its tunnel and gnaw into the wood: gnaw, gnaw, gnaw—pause, gnaw, gnaw, gnaw—pause, is the pattern which characterizes its life in the wood for from one to five years. The grubs tend to burrow along the grain of the wood; that would be up or down the tree were it standing in the forest, but sometimes they do move across the grain, either nearer to, or further from, the centre, till finally the wood is honeycombed with galleries and is completely rotten. The larvae move across the grain—from one ring to another of the wood—for various reasons, to avoid another larva approaching in the same fibre and to find places where the food is more to their taste—literally so. In particular, the grub must obtain a supply of protein in order to add to its own muscles and thus continue boring and also to be a reserve for itself when it comes to pupation and maturity, for a considerable stock of protein must be incorporated in the next generation of eggs. Its carbohydrate is easily obtained from the actual wood itself, as has already been explained, for this insect can digest cellulose directly; nevertheless, sugar is an easier material to feed on and the grub grows more quickly where the wood contains a supply of sugar, and always seeks out this material in its slow travels. Sugar occurs in the cambium and tissues of some woods.

The grub also needs oxygen, which can only be obtained from the air by diffusion through the wood and the galleries; the position of a gallery is a compromise between the ease of obtaining oxygen and food, and of competition with other grubs for these same items. It is obvious that a short gallery or a gallery near the surface will more readily obtain oxygen than a long or deep penetration. In spite of being able to digest cellulose the grub eats far more wood than it needs for its carbohydrate supply alone, because it wants other materials as well (amino acids, proteins and so on); consequently, a great deal of waste is excreted behind the grub and this fills the passages through which they have passed. *Anobium* leaves a fine granular powder, with little cylindrical pellets in it. This dust, or frass, in the gallery means that oxygen diffusion down the gallery is slow and so is the dispersal of the carbon dioxide generated as the grub breathes.

The action of burrowing calls for the use of a lot of energy, as any miner will tell you, and consequently the grub, as it bites forward, uses up the supply of oxygen in its immediate neighbourhood, for as the insect breathes, it burns up some of the carbohydrate in its body and turns it to carbon dioxide and water, in the same way as man or any animal. Insects can usually tolerate a considerable volume of carbon dioxide gas, but at some point the grub must wait for more oxygen to seep in, either through the burrow or through the wood itself. The life-cycle of *Anobium* is much quicker in the forest than in the house, perhaps because the oxygen is changed more quickly in the former situation, for the temperature difference between night and day is greater out of doors than inside, which means a bigger expansion and contraction of the air in the galleries of the borers and a bigger exchange of the gases in them. Similarly, the wind blowing in the forest means more tiny variations of pressure outside the gallery which again facilitates gas exchange.

On the other hand, the presence of the mass of frass in the gallery prevents the access of parasites, such as certain mites and other insects, which might prey on the grub. Only at the last moment before emergence are the grubs likely to fall victims to parasites. There is a very delicate balance between success and failure, and by no means all the grubs that hatch and start their galleries into the wood are successful; some die of oxygen starvation, some from lack of protein and some from poisoning.

The poisoning needs explaining. The only part of the wood which is living is the sapwood, and this transmits the water and salts extracted from the soil to all parts of the living tree. The heartwood of the tree is dead—it is frequently a different colour—and it serves to support the tree and carry it ever upward so that the leaves can get the light and tower over other leaves in the eternal struggle for existence, which is just as intense among plants as among animals. Trees have waste products of their metabolism and they dispose of these to two places—the bark and the heartwood. Quinine and camphor are examples of waste products found in bark and wood which are valued by man, and some of these substances are of use to trees in preventing the attack of enemies. The oak excretes a certain amount of tannin into the heartwood, though most of it goes to the bark, and if some *Anobium* grubs get into a pocket of tannin and are not able to extricate themselves in time they just wither away.

By August 1555, when Richard Barton gave his feast to celebrate the erection of the ridgeboard, there were roughly a hundred furniture beetle grubs boring along the old ship's timbers which formed part of the stairway in the house and some of the uprights in the main bedroom wall. The house was now but a frame, and soon John was at work directing almost unskilled labour in the filling in of the framework with walling. Some bricks left over from the chimney-stack were used to make a footing, while the rest was completed with wattle and clay daub and then plastered with lime mortar; finally all was white-washed. Some of the older split hazel, which was used for filling in the bigger spaces of the timber frame, had *Anobium* eggs laid on it and the grubs started their galleries in this in much the same way as they did in the old timber. Basketwork and hurdles reach a condition attractive to *Anobium* much more quickly than do large timbers of oak. They were unfortunate in that the clay daub and plaster were so well applied that the air diffusion through the wall was not enough to keep the grubs alive for long and they had all died by the summer of the following year.

This then was the position when John and Mary were married in March 1556. There were ambrosia beetles, or pinhole borers, both adults and larvae, boring in the new oak beams. As the green timber dried the wood became unattractive to the beetle and was no more attacked, so that the house was free of pinhole borers by March 1557.

Next there were the powder post beetles, or *Lyctus*. These were not very numerous because they only attacked sapwood and not much of this went into the building of the house. The starch in the sapwood was soon used up and the wood became of no more use to the *Lyctus*. The house was free of them by 1560.

The furniture, or *Anobium* beetle, was a different matter, for, coming in on the wattle and the old ship's timbers, the house was never again free from it. In fact it might be said that Bartons End belonged as much to the *Anobium* as to the humans who lived there.

Even so, a still more deadly enemy was waiting in the forest—the death watch beetle. Each year it appeared to send out scouts to see if the house was yet ready for it—'appeared' because these scouts were only part of the annual march of adults looking for new territory, most of whom perished. It was nearly a hundred years before the death watch beetle established itself; in order to follow the subsequent history of these insects we must first briefly consider the history of the house in the next four hundred years.

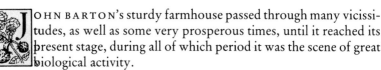

Chapter 4

Bartons End

JOHN BARTON's sturdy farmhouse passed through many vicissitudes, as well as some very prosperous times, until it reached its present stage, during all of which period it was the scene of great biological activity.

The house supported a considerable collection of animals, nearly all of whom depended to some extent on the humans, and we can best follow the ramifications of this life by studying a schedule of the people who owned or rented the house from 1555, the year it was built, till 1984. This is shown in tabular form on pages 8 to 11.

Three generations of Bartons lived in the house, but in the fourth there was no male heir, and in 1631 it passed to Mark Barton's daughter, Elizabeth, who had married a young farmer, Robert Onway. Elizabeth Onway's son Mark, who inherited the property in 1660 and only lived five more years, found the old-fashioned system of bedrooms opening one within the other very irksome and at last the rather pretentious gallery to the hall could be put to some use, for it was comparatively easy to open doors from these rooms to the gallery. Elizabeth lived to a ripe old age (she died at the age of eighty in 1688) and adored her grandson James, whose father, Mark, had died of the plague when the boy was only thirteen years old. As a consequence James was rather spoilt by his grandmother, went to London, now a very gay spot after the restoration of Charles II, and lost a great deal of money at Hazard. He was obliged to sell the property as soon as the old lady died, in order to pay his debts.

The purchaser was Joseph Munroyd, a retired merchant in the West India sugar trade, who had long desired to own a property in Kent. He made Bartons End much more elegant by replacing the thatch with tiles, an action which had a considerable effect on the life in the house, and he put ceilings below the floor joists of the upstairs rooms over the hall and the parlour. It was a wise move on his part as it enabled his grandson, Charles, to marry the wealthy Susannah Aylesford, who, he thought, certainly would never have consented to live in a thatched house. A large number of improvements were made to the property when Susannah arrived. Ceilings were put in all the rooms and fireplaces constructed at the south end and in the main bedroom. The ceilings to the bedrooms meant that a loft was available for storage purposes above them, making a new environment with a particular life of its own. It was at this time that a pump was installed in the kitchen and the well outside the front door was covered in; another pump was also placed in the farmyard.

Charles and Susannah's eldest son, George, inherited Bartons End together with a substantial sum of money, which enabled him and his Hanoverian wife further to embellish the house. They blocked up the immense fireplace in the hall and put in a small coal-burning stove, and this encouraged the firebrats who found this new arrangement much to their advantage. George and Sophia had only one child, a daughter, who inherited the property in 1808. She was consumptive, steadfastly refused all offers of marriage and though comparatively young, ran the farm with quiet efficiency, helped by her bailiff, Luke Burrows, until her early death in 1817. The heir, her first cousin, sold the property to a hero of the Penninsular and Waterloo, Major John Hackshaw of the 2nd Light Dragoons. Both the major and his son William were devoted to Bartons End and considerably improved the farm and the house. They installed a water tank in the roof, fed from a hand-worked force-pump just outside the back door, and a pipe from this tank led to taps on the landing and in the kitchen. The Hackshaws were very proud of these modern innovations which greatly lightened the labours of the servants and much facilitated cleaning the house. This in turn had an influence on its biology; during this period the bed-bug vanished and the flea was but a rare visitor.

The major enlarged the cellar and installed a considerable quantity of port wine there, his share of the supply which had been showered on the regiment by the grateful inhabitants of Portugal. Again this action influenced the life in the house, as we shall see later.

William Hackshaw's eldest son, Alfred, was a great disappointment to his father; he disliked both farming and England and eked out a dilettante existence in Venice from the remains of his grandfather's legacy and by letting the farm. He never married. His brother David was killed in the Crimea. Alfred died in 1898 and left the property to his nephew, David's son, Oscar Hackshaw, who was in the Indian Civil Service. Oscar sold the farm lands and meant to retire to Bartons End with the modest fortune he had acquired in the East, but when he actually saw the farmhouse, which was in bad repair after two tenancies, on a raw January day as it happened, he refused to have anything more to do with it. The house was too big to let without the farm attached and in too poor a condition to sell: moreover, Oscar had so antagonized his solicitors with his fiery temper that they made little effort to dispose of the property for him.

Bartons End remained untenanted by humans for ten years, apart from the occasional tramp who passed a night or so there, and the life in the house was much changed. There was still life in it—but with quite a different biological balance.

In the early summer of 1908, Robert Dunchester and his Welsh wife, Myfanwy, were making a bicycle tour of Kent. He was a moderately successful portrait painter with a small private income. One evening as they cycled down a remote lane on the edge of the Weald they came across the decaying Bartons End lit by the setting sun. Tired and thirsty they asked Sarah Claridge, the caretaker standing at the gate, for a glass of water. Not suspecting the bicyclists of being potential buyers of the property (for no caretaker wants to be ousted from a comfortable place and in any case would madcap bicyclists be the sort of people who would buy an old house?) Sarah gave them some goats' milk and chatted away. The charm of the old house grew on the couple and, despite all Sarah could do to denigrate the property, the Dunchesters eventually bought the house for a very small sum.

It was the heyday of the week-end in the country and the Dunchesters were very much 'taken up' by society in their work of saving and restoring the old house. In fact Robert never escaped from his portraits, because the demand to be painted with a somewhat Tudor back-ground—the 'period portrait' which he developed—kept him very busy and even today a few people will remember 'The Riding Lesson' (Lady Mary Feld about to mount a horse outside Bartons End) and 'Mrs J. K. Schultz' (the wife of the Congo copper magnate, rocking a Tudor cradle with her foot, in the hall at Bartons End) which appeared in the Royal Academy exhibition of 1914.

The house was much altered; a studio with a roof-light was construc-ted at the north end; a bathroom was installed, also a complete hot water system and finally central heating. The kitchen was removed to the north end of the building, below the studio; the old beams in the ceiling were all uncovered again and also the huge old fireplace in the hall; the cesspit sewage system was added and electric lighting was run throughout, at first from a home generator and finally from the mains. The general effect of all this was greatly to discourage the cockroaches and greatly encourage the woodworms, whose presence was not realized until a flying bomb fell and exploded near the house in 1944. The explosion loosened part of the roof, broke some windows and violently shook the whole structure, but did no fundamental damage to the house, which might almost have been constructed to withstand such an attack. The heavy oak mortised beams had a give and spring in them lacking in steel or brick. The filling between the upright beams could be sucked out with little damage to the real structure and moreover was easy to replace; lead light windows bend in and out to a much greater extent than do large sheets of glass and so were less damaged. The house was very vulnerable to fire, but could resist any ordinary bomb that was not a direct hit. This event released clouds of wood-borer dust, and drew attention to this enemy within. It led to an action causing a biological change as great as that caused by leaving the house empty in the early 1900s—all the woodwork was treated with insecticide.

Robert F. Dunchester (b. 1910) died in 1948, leaving the property to his wife for her lifetime, with a reversion on her death, or remarriage. It was then to go to his nephew and two nieces, Phillip, Caroline and Myfanwy Frazer. As plans were being made to divide the house into two unequal portions, the larger for the Frazers and the smaller for Louise, the young widow announced she was going to marry again and that the property should go to the Frazers. This considerably eased the situation as it was proving very difficult to plan the division of the house.

The Frazer children, in order to preserve the house in the family, decided to turn the place into an old people's home. To do this they formed a company, Frazer Properties Ltd, and appointed their mother, Sloane Frazer, still comparatively young, as managing director.

Extensive additions and alterations were made to the outbuildings but though the usage of the rooms in the old house was altered, the actual structure was hardly touched. The central heating was improved, care being taken to make the pipe runs, where they passed through walls, insect and mouse proof. The old dairy and store became the kitchen and scullery and the parlour was turned into cloak rooms and a bar. Upstairs there were sitting rooms, management offices and a pantry where elaborate teas were prepared, the patients having the choice of Indian, China or comfrey brews. A store room was opened up in the loft and the roof was insulated against heat loss with glass fibre. Although double windows were not fitted, air passages to the fireplaces were made, and draughts thus prevented.

To comply with the fire regulations, an external iron staircase was constructed on the north side of the house as a fire escape, up which the Russian vine (*Polygonum baldschuanicum*), an unusual kind of Virginia creeper, was trained to conceal it, possibly in defiance of the said regulations. Bedrooms were made in the converted stables and cow-sheds. The big Tudor barn became an assembly hall and the oast house a library and reading room.

Phillip Frazer, a young man of twenty-eight at the time of the company's formation, had become a solicitor and looked after its legal aspects. Caroline Frazer had studied medicine and eventually became the new company's resident medical attendant, while Myfanwy II Frazer, who had studied horticulture at Wye College, became the new enterprise's head gardener.

In 1962 an American girl named Desirée Barton called at Bartons End, seeking information on her probable English ancestors. Phillip Frazer offered to help her; they fell in love and were married that year. They lived in a nearby house.

Myfanwy Frazer married a local farmer, Giles Thompson, in 1967 and has three children, Phillip, Caroline and Sloane Myfanwy III.

The biological alterations made to the house by the change of use were not dramatic. The biomass varied in quantity according to the time of day and the weather. Among the different kinds of animal life still in the house the humans were the greatest weight. At meal times there was a lot of biomass there: at night, none. During wet weather there was more than during fine. Certain non-human creatures still influenced the life of the house. For instance, the Dutch elm disease from 1975 onwards was steadily killing all the elms in the neighbourhood and destroyed most of the fine elm avenue leading to the front door. Three species of bark beetle, *Scolytus multistriatus*, *S. scolytus* and *S. laevis*, burrow beneath elm bark. They carry a fungus with them which breaks down the elm wood already chewed away by the beetles, enabling the insects to use the material as food. Unfortunately, in Canada the fungus, *Ceratocystis*, developed a vigorous, new strain which did more than its simple task of providing food for the *Scolytus* beetles; once in a burrow it grew on and on into the elm tissues, blocked the sieve tubes and water channels of the trees and killed them. Eventually the Canadian strain reached Europe in insufficiently barked timber, and it is now killing elms over the whole continent. It was first studied in the Netherlands, hence its name, the Dutch elm disease.

One result of the new fungus's activities was the creation of a new industry—the provision of wood-burning stoves and of logs to feed them—for the dead and dying elms were of use for little else. Consider-able numbers of *Scolytus* beetles came into the house as the splendid elms were converted to firewood and the insects perished in the big open fire of the dining room and the wood-burning stoves of many of the other rooms. It remains to be seen whether, in say ten years' time, when all the dead elm will have been consumed, if the wood-burners can still be supplied with fuel or if they will then be torn out. Maybe that dangerous combination—*Scolytus* and *Ceratocystis*—can transfer its activities to another species of tree, perhaps to the oaks, and the cycle repeated, or perhaps a strain of elm may arise able to resist the fungus, or some other control be found. There are two promising leads.

Dr R. J. Schaffer, in the Netherlands, has recently found that elms deliberately infected with certain bacteria (Pseudomonads) are able to resist the attacks of the fungus.* Another worker in this field, Dr C. M.

* Schaffer, R. J. 'Biological control of the Dutch elm disease by *Pseudomonas* species'.

Brasier of the Forest Research Station, Alice Holt, has discovered that a disease of the fungus exists which reduces the ability of the virulent strains of *Ceratocystis* to damage elms†. Perhaps the elms will come back to Bartons End after all.

The alterations made to the old house meant a certain amount of reconstruction. All the timbers used for such work had been pre-treated with insecticide and the existing rafters, laths and beams, which had been treated with insecticide in 1947, were resprayed in 1978, so the attempts of various new wood-boring beetles to establish colonies were of no use. For example, the large and handsome Longicorn, *Hylotropus bajulus*, by 1978 had spread down from Surrey to Kent and laid eggs in softwoods. In the treated timbers they either did not hatch or the grub died soon after hatching, but they did develop in some untreated bits of whitewood furniture. The larvae spent a year or so in the wood, pupated and then emerged as adults from largish, oval holes. Often this furniture was just a veneer of apparently sound wood over a mass of tunnels and frass. Such attacked pieces usually ended up on the big, open dining-room fire.

The furniture beetle (*Anobium punctatum*) pursued a similar policy and found the wickerwork chairs that many of the elderly patients thought so comfortable to be ideal spots for their eggs. Many a favourite chair collapsed in a pile of bits and dust, and Sloane Dunchester insisted that any wicker furniture or storage baskets brought in by the patients had to be treated with insecticide.

The new Bartons End still had a certain amount of non-human life in it besides the *Scolytus* and furniture beetles. These animals were mostly visitors, because the house had been so carefully protected with chemical defences and ever more effective vacuum cleaners that none of them was able to establish a bridgehead for any length of time.

Of these visitors the most welcome were the house-martins and the least so the flies and wasps, which are mutually antagonistic. 1982 was a poor year for wasps, one of the results of which was a large number of flies, particularly the tiny vinegar fly, *Drosophila*.

Annals of Applied Biology (1983) 103. 21–30.
† Brasier, C. M. 'A cytoplasmically transmitted disease of *Ceratocystis ulmi*'. *Nature* 305. (15 Sept 1983) 221–223.

Wall to wall carpeting, where it was of wool and against the skirting boards, was sometimes found to be infested with the woolly bear or carpet beetle, because the vacuum cleaner could not exercise its full force right up to the wall. But such invasions did not last for long; insecticidal sprays or dust soon put an end to them. The final accolade was the discovery of some old timbers showing the exit holes of the death watch beetle. 'To have the death watch beetle today,' said Dr Dhusti, a patient and a retired Indian scientist, 'is a status symbol.' The wood had been treated and there were no live beetles in it, because, of course, it only occurs in old timbers, of which there are now not very many.

Desirée Frazer was proud of the Barton name and pleased to have found the old house, which she considered to be her family's original home: she persuaded Phillip at the time of their marriage to change his name to Frazer-Barton. She frequently visited the old house and planned gradually to eliminate the old people by means of natural wastage and to live there eventually with Phillip and a host of their happy children. Alas! she has none so far and she is forty-three. However, she consoles herself with the thought that the old house is providing shelter, comfort and beauty for a number of elderly people, who still have something to give.

> 'For out of olde feldes, as men seith,
> Cometh al this newe corn froe yeer to yeer;
> And out of olde bokes, in good feith,
> Cometh al this newe science that men lere.'

G. Chaucer. *The Parlement of Foules*

✳ *Chapter 5* ✳

More Wood-borers

AT THE end of Chapter Three I mentioned a dangerous insect that appeared to be waiting in the woods till Bartons End was ready for attack. This was the death watch beetle, which cannot attack living wood but must live in hardwoods in a certain state of maturity. There were but few of them in the forests around Ashwell, for the young larvae could not penetrate bark; consequently, they only flourished in those hollow trees and branches which had been decaying for some ten or twenty years. If wood has been weakened by the growth of a fungus in it, as frequently happens with fallen trees in the forest, the death watch beetle larvae are able to attack it more readily. The 'beef steak fungus', a large reddish bracket growth often found on old oaks, gives a valued staining to the wood and is good to eat but also pre-disposes the timber to the attack of the beetle.

In a Tudor building, on the other hand, there was a good supply of sawn timber with plenty of surfaces for egg-laying, and it would be at the right stage of maturity when it was sixty to a hundred years old.

The adult death watch beetles do not fly very much, in contrast to the *Anobium* we have already discussed, and after pairing, the females usually lay eggs on the same wood from which they have emerged. If there is no suitable site they walk off looking for one and frequently settle on the ends of young fallen branches. The adult beetles are quite large (from a quarter to a third of an inch in length), almost twice the length of the *Anobium* and rather squarer in shape. They are red-brown in colour with tufts of short golden hairs.

The adult beetles are ready to emerge from their pupal cells in May or June. Just before and after they emerge, they make their characteristic tapping noise, very like the ticking of a watch, by striking their heavily armoured heads down on the wood. It is the signal by which the sexes find each other; seven or eight sharp taps are heard per second, then all is still, then the sound is repeated, perhaps from a different though near-by position, and this will be another beetle answering the call.

It is an eerie sound when heard late at night, so that it is not surprising that the tapping of the beetles was said to predict the death of Elizabeth Onway, when she was taken ill in 1688. As soon as an old family servant Hannah heard it, sitting late one night at her mistress's bedside, she sent to London for James Onway, for she knew the old lady was going to die. It was not surprising that Hannah proved right since Mrs Onway was so very old. In point of fact, if a house has the death watch beetle in it the tapping can be heard in any May or June, but it is a faint sound and only heard when all else is still—as it will be when one is watching a sick person, or a body: the name—death watch beetle—is double-edged because it can mean both that the sound is heard during the death watch and also that the noise is like the ticking of a watch.

The adult beetles appear in June and find each other by means of the tapping. The females lay their whitish, slightly pointed eggs, some sixty in number, in clusters of three or four, on the surface of rough wood usually near the old exit holes of the adults, which are oval in shape and much larger than the round exit holes of the furniture beetle (the *Anobium*). In two weeks the eggs hatch and, unlike the *Anobium*, the larvae at first crawl over the surface of the wood until they find a suitable crack in which to start boring; very frequently the larvae make their entrance through an old exit hole of the adults. The grubs tunnel in the wood in much the same way as do the furniture beetle larvae, though the galleries are bigger and they leave a different sort of frass behind them; that of the death watch beetle has oval pellets in it much bigger than those left by the furniture beetle.

The grubs occasionally live in the wood considerably longer than those of the furniture beetle; two or three years is common and up to ten years has been known. The period will be regulated by the temperature,

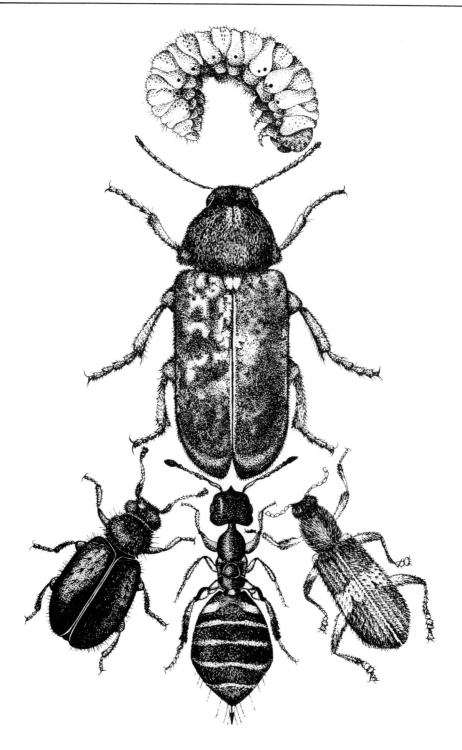

Death-watch beetle and larva, and its enemies, black Clerid beetle,
Theocolax *and soft brown beetle*

and the availability of oxygen, water and food. The grubs have an exceptional gut activity for they can turn the cellulose of the wood to glucose and they like wood that has been attacked by a fungus because the mechanical weakening of the tissues makes it easier for them to bore their galleries. As the grubs reach maturity they seek more oxygen and move towards the surface of the wood again, where just beneath the surface they construct their pupal cells: they become adult in the late autumn, but pass the whole winter quiescent, resting and waiting for the right moment—May and June—to complete the very different remainder of their life-cycle.

The death watch beetle was continually being brought into Bartons End on the firewood, and every June a few adults would emerge but at first laid no eggs on the house timber, though they tested it, because it was not in a sufficiently attractive state until about 1660, when some females emerging from the wood pile laid their eggs on the open floor joists of the hall. A new form of life was now established in the house.

We can now see how the wood-borers affected the house as a whole. There were three centres of *Anobium* infestation: one spreading from the original attack on the old timber which was built into the staircase and gallery, thus dating from the building of the house in 1555, one developing in the floor joists and underside of the ceiling to the hall and parlour, which came from insects brought in with the firewood, and one growing from a small infestation of winged adults coming from the forest and getting to the rafters under the eaves of the thatch. But these last two centres were not able to form colonies until 1605, when the original green wood in the main structure of the house became sufficiently attractive for the females to lay eggs on it and for the resulting larvae to survive. The beetle also attacked the unpolished surfaces of the old furniture which the original young couple had been given. It did not make very rapid progress; in the first place a number of parasites and predators—in the form of other insects—kept it down, and in the second case only the old wood in the house was palatable. By the time the new wood was fifty years old the furniture beetle began to thrive and the old matriarch, Elizabeth Onway, frequently chid her maids for being idle ne'er-do-wells when she found the dust from the exit holes lying on the floors or furniture.

In the house the furniture beetle was considerably checked by the installation of the fashionable ceilings in the first half of the eighteenth century; indeed they may have saved the house. Laths were nailed to the beams and the ceiling then plastered. This meant that although the larvae in the beams could continue to live, they did so at a slower rate as it was more difficult for them to get oxygen, and most of the adults could not get out to the open air when they had pupated and their turn came to take up their free life. Adult furniture beetles have a great capacity for boring: they have been known to penetrate copper sheet, or lead and some of them showed great persistence when they were trapped under the plaster at Bartons End, for they tunnelled their way out. But even so it was a barren victory, because there was no convienient wood surface for egg-laying. Some of the adults came out as usual into the air space between the joists, now dark and enclosed by the ceiling, where they were not able to breed: yet others penetrated right through the floor boards, and came out to the open air that way. There was little success in this direction, for Anne Munroyd was very proud of her house and polished all her floors with beeswax, which had three effects on the beetles. It stopped their absorbing oxygen and discouraged the movement of the grubs to the floor surface; it actually killed some beetles as they attempted to get through the wax layer, but most important of all it made all the floors unsuitable for egg-laying.

Nearly all the furniture beetles that did escape from the catastrophe of the plaster ceiling were obliged to lay eggs on the unpolished parts of the furniture, on the roof rafters, or on the firewood, this last being no way to secure the future of the species. The population would have declined still more had not Joseph Munroyd replaced the thatch with tiles (1690), which meant that a large number of tiling battens had to be fixed across the rafters, the tiles being secured with wooden pegs. These enabled the furniture beetle to keep up its numbers, though not to destroy the house, as it would have done had there been no plaster ceilings.

The bedrooms were fitted with ceilings and a loft made to the house in 1752, and in this new arena a struggle for supremacy appeared to be going on between the furniture beetle, the death watch beetle, other insects, spiders and eventually some bats. As the rafters aged the furniture beetles found them less attractive, and the death watch beetles more so, particularly where damp conditions had allowed fungi to invade the wood tissues. Nevertheless, the very success of these insects,

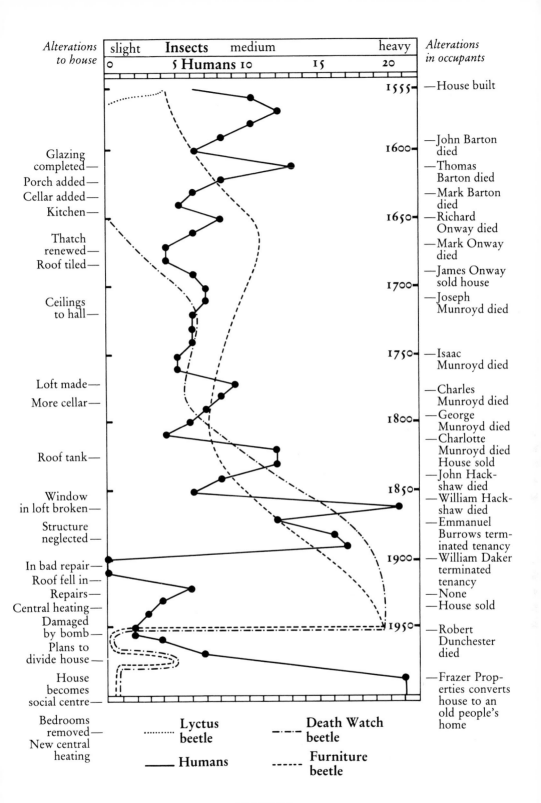

Alterations to house

slight · **Insects** · medium · heavy

5 **Humans** 10 · 15 · 20

Glazing completed—
Porch added—
Cellar added—
Kitchen—

Thatch renewed—
Roof tiled—

Ceilings to hall—

Loft made—
More cellar—

Roof tank—

Window in loft broken—

Structure neglected—

In bad repair—
Roof fell in—
Repairs—
Central heating—
Damaged by bomb—
Plans to divide house—

House becomes social centre—

Bedrooms removed—
New central heating

Alterations in occupants

1555 —House built

1600 —John Barton died
—Thomas Barton died
—Mark Barton died
1650 —Richard Onway died
—Mark Onway died
—James Onway sold house
1700 —Joseph Munroyd died

1750 —Isaac Munroyd died

—Charles Munroyd died
1800 —George Munroyd died
—Charlotte Munroyd died House sold
—John Hackshaw died
1850 —William Hackshaw died
—Emmanuel Burrows terminated tenancy
—William Daker terminated tenancy
1900 —None
—House sold
1950 —Robert Dunchester died

—Frazer Properties converts house to an old people's home

.... **Lyctus** beetle   –·–·– **Death Watch** beetle

——— **Humans**   - - - - **Furniture** beetle

FIGURE I

Population of humans and wood-boring beetles at Bartons End

after a certain point, tended to limit their numbers. When the surface of a beam had become honeycombed with galleries full of frass it was difficult for the young larvae to penetrate the mass of waste or useless material and many died before they could reach food; even some adults died as they tried to tunnel their way out and many more used up much of their food reserves and so laid fewer eggs when they did emerge.

The wood-borers were not without insect enemies: there were two species of small Clerid beetles—the black Clerid and the soft brown—the larvae of which, in spite of the mass of frass in the burrows, penetrated down them and devoured many of the grubs, thus keeping down their numbers. But their greatest enemies were the ichneumon flies, or their near relatives. The ant-like *Theocolax (Pteromalidae)* was the first hymenopterous parasite to establish itself in the house. These small wasps ran in and out of the holes of the furniture beetle larvae and when the females found suitable grubs they would lay eggs on their backs, but only one to each grub: the parasites grew as the grubs grew and eventually killed a considerable number of them. *Theocolax*, the flatterer of the Gods, was present in the house until the virtual extinction of the furniture beetle, but it became of much less importance in keeping down the numbers of the beetles than did a tiny fly-like creature, *Spathius*, two species of which did much damage to the furniture beetle grubs as they approached the surface of the wood to pupate. *Spathius* belongs to the Braconid family and looks like a small fly, though it is really wingless. The *Spathius* females walk over the surface of the wood and know when they are over a gallery containing a suitable *Anobium* grub. The *Spathius* holds out her antennae straight before her head and parallel to the ground and can thus catch the vibrations set up by the gnawing of the host grub beneath. When she has found a suitable spot the *Spathius* female bores down with her ovipositor through the wood and leaves an egg on the furniture beetle grub below, and so the usual parasitic cycle continues.

As the woodworm in the house increased so did the numbers of *Spathius* insects; these, the Clerid beetles and *Theocolax* did much to preserve the house, for at the height of their population, as much as sixty-five per cent of the wood-borers were attacked by parasites; but this still left thirty-five per cent of the insects slowly eating away the timber, and it is rare indeed that parasites completely destroy an infestation, for to do so—to be too successful—would be to ensure the parasite's own extinction. Thus the law of nature—the survival of the fittest—sees to it that parasites and predators usually have some alternative host, or else are not too successful. We do not know those species of parasites which, having no alternative host, have been completely successful in extinguishing their hosts because they have become extinct themselves as well. At the peak of the attack, in May 1939, there was such a large emergence of *Spathius*, because woodworm attack in the roof was getting serious at this time, that people complained of a 'plague of small flies' and set about them with fly-spray, an action which helped maintain the population peak of beetles at a high level until the wood was treated with insecticide in 1947.

The death watch beetles so weakened the roof at the north end of the house that it fell during a winter storm in 1903 and this gave access to bats, who eventually established a fair-sized colony in the loft. The bats consumed considerable numbers of both the furniture and the death watch beetles and this is another example of the dangers of too great a success.

The spiders also played their part in keeping down the numbers of beetles, but they will be dealt with in another chapter. The humans did not realize that the woodworms had brought the house into such a bad state until it was bombed in 1944. After the bombing the roof was repaired temporarily with corrugated iron. The whole house was treated with a gamma benzene hexachloride insecticide dissolved in oil in 1947, which very nearly exterminated all the woodworms in the house. It did not destroy them completely because the insecticide could not penetrate far into the wood where the deepest seated grubs burrowed, though these insects would surely die as they sought the surface to pupate: in addition there were still a number of furniture beetles, particularly in the furniture made of plywood, and in basket-work about the house. None of the adults that emerged from furniture, kindling wood and basket-work was successful in laying viable eggs on the structure of the house, for the insecticide either killed the females as they traversed the beams or the larvae as they hatched. The death watch beetle was extinct in the structure by 1955 though the purchase of an antique dresser reintroduced it for a while in 1958; by which time the strange ghost-like tapping could again frighten the lone, sad watcher in the night. It was thought that the house timbers would be safe until about 1975 as by that time the insecticide put on in 1947 might have lost most of its power. But the Dunchesters had reckoned without the threat imposed by plywood furniture. Such furniture is rarely made with wood treated with insecticide, although it is comparatively easy to incorporate it in the glue used in making the plywood. To do so is an additional cost and such furniture is always sold as cheaply as possible.

Sloane Dunchester bought some plywood tables and chests of drawers in 1975 to furnish some of the patients' bedrooms. By 1977 some of them were showing considerable attacks of the furniture beetle and she realized that the purchase had been a false economy. Most of the infested furniture was burnt: this event reminded her that the twenty-five-year guarantee on the house beams had expired, a point the pest control salesmen were quick to press. During the summer of 1978 all the beams and roof woodwork were again treated with insecticide, this time guaranteed for forty years. It made the family reflect who among them would live to see this guarantee expire? Possibly all the Thompson children, who, in 2018, would be in their late forties. Even Myfanwy Thompson herself might be there, aged seventy-six.

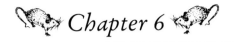

Large and Small Mammals

THE BIOLOGICAL activity of the mammals at Bartons End was dominated by the humans, some of whose social activities are part of this chapter, and were also described in Chapter Four: indeed the social and biological activities impinge one on the other. A certain social pattern among some of the hymenopterous insects has enabled a large environment to be occupied by vast hordes of ants, bees and wasps. The social pattern of a pride of lions enables them to co-operate in obtaining their kills and caring for the young, while social co-operation among man has led him to dominate on earth, at any rate from soon after his appearance until the present. There is no need to describe the natural history of the more familiar mammals—man, the cat, the dog, in such detail as I have described the less well known insects, but we do want to think of these animals from the point of view of numbers, descent, heredity, how and why they got to the house and why those that left it did so.

The mammals in Bartons End were never completely driven out because, though there were ten years from 1899 when there were no humans present, which meant that the place was also abandoned by mice, cats, and dogs, yet the loft became inhabited by two species of bats. Consequently mammals, as a class, have as long and uninterrupted a history as the wood-boring beetles.

It was natural that humans should dominate life in the house, because they had built it for their own purposes. The reason why Richard Barton was able to build it in the first place was that human technical ability took a big step forward during the times of the Tudors. A better understanding of the principles of agriculture led to better crops, which in their turn meant the release of more hands and brains for activities other than gathering or getting food. The number of mouths a man could feed by his efforts on the farm was constantly increasing. In the Stone Age the whole population was engaged in getting food by plant-gathering, agriculture or hunting: it took the activity of one man to feed a man and his family. Gradually skilled flint-workers set up what were in effect factories for making axes and arrowheads, which they traded for food. The principles of Adam Smith's *Wealth of Nations* were being applied: it was mutually advantageous to the axe-maker and the axe-user to exchange axes for food and it is indeed surprising to discover over what long distances men would travel in Neolithic times in order to get a well-made stone axe.

As time went on, the methods of agriculture, fishing and hunting improved so that fewer and fewer men needed to be engaged in it—the husbandman (using the word to include all forms of food-winners), who could at first feed only himself and his family became able to feed several people, which left a surplus of hands, and brains, to devote to other economic and social activities—to smelt metals, build houses, invent the wheel and money, appoint chieftains, kings, priests, to learn to write, to calculate, to paint and make music, but above all to develop nations, armies, governments, war and in fact the modern world. In late mediaeval times about three-quarters of the population were engaged in food-winning, which meant that the husbandman could feed about one and one-third persons, but by the time Elizabeth I had come to the throne agriculture and industry had so improved that each husbandman was supporting perhaps three members of the population beside his family, and by the end of the century (1599) the figure was about one to four. Today in Britain 1,200,000 farmers and farm workers supply forty-nine million people with half their food, which is equivalent to one man providing twenty-one people with all their food.

The improvement was due to better management of the land, made possible by enclosure of the common fields. The Bartons went ahead particularly fast because they had to lie low during the Marian persecutions, and so turned their energies away from politics towards their own land.

From the time Bartons End was built (1555) to the present day the number of humans in it varied greatly—from zero to twenty-one. When John and Mary Barton took up residence on the day they were married there were six of them, for in addition to the newly-weds there were also two maids and the two old farm labourers. Children were born to them—some died and some throve—maids and labourers came and went, parents and relations resided there; which caused the numbers of humans to fluctuate around the average of nine. This can be seen by the diagram on page 55 which shows a line drawn to represent my census of humans taken in 1556, and 1560 and at every ten years from that day to 1980. The zero point was reached in 1900, when the house had been empty a year, because Oscar Hackshaw refused to live there himself and the house could not be sold. The peak (21) was registered shortly before this (1860) during the tenancy of Emmanuel Burrows who, in addition to having a large number of children, also boarded various brothers and cousins, who helped him work the farm. Bartons End is now an old people's home and the question of how many people it contains depends on the time of day the count is made. The old house is the central meeting place for the patients and staff. There would be some forty souls there on a wet afternoon and none at all at night.

The fertility of the humans in the house varied as much as did their numbers; it ran from zero—Charlotte Munroyd (b. 1786) and Louise Dunchester (b. 1916) to twelve—Emmanuel Burrows (the son of Charlotte's bailiff) and his wife, who were tenants from 1853 to 1869. Charlotte was consumptive, unmarried and died young (thirty-one years): and Louise Dunchester has been married twice over, some forty-eight years, and has had no children.

These were the numbers that resided there. On special occasions, such as parties, weddings, and funerals many more humans were found, temporarily, in the house. On 1 June 1584, when Thomas Barton married, there was a great gathering of the Barton and Standish families and their servants so that some sixty-five persons were present. This was greatly exceeded by the reception given on the occasion of Myfanwy Frazer's wedding in June 1967. About 220 people, including guests and staff, assembled, weighing some twelve tons. A marquee had been erected in the garden so that not all of the 220 were ever in the house at the same time. Even so, a good proportion of them were at times, say a little over half the above number. It was quite a strain on the old flooring. The party was a great success as the somewhat diverse worlds of publishers, agents and authors mixed with the county one of farmers and gardeners, so that a biologist might be excused for thinking of them as distinct sub-species and wondering if they would prove inter-sterile, a concept in fact disproved in time by one of the literary girls there meeting, and later marrying, a local hop-farmer and producing a long line of children.

The 120 or so of individual humans is a small number compared with that of some of the mites and insects found there; even other mammals, such as mice and bats, sometimes exceeded this figure.

The house did not descend from father to child in any one family for long: some of the original gene pattern of John and Mary Barton, which was stamped on their children by laws and chances of heredity, passed through five generations before the Barton descent disappeared altogether, by the sale of the house in 1689. The Barton name was lost in the third generation, when Elizabeth Barton married Richard Onway, but this made no difference to the genes, or Barton 'blood' as we would say in more familiar terms. Five generations from 1556 to 1689 meant that the gene pattern had been halved five times so that James Onway and his sister only carried $\frac{1}{32}$ of their genes from the original owner of Bartons End (John Barton) and another $\frac{1}{32}$ from his wife (Mary Halstead) which makes $\frac{1}{16}$ of the genes from the two of them, or $6\frac{1}{4}$ per cent. However, as the generations extend one after the other, lines of descent tend to mingle by intermarriage, though the individuals may well be unaware of any relationship, and thus some of the original genes will be brought back into the family again.

Any discussion of the gene make-up of the humans depends on the point from which we start, which in this case is the first man and wife in the house, but of course these two people were themselves the result of a complex descent stretching back thousands of generations.

Mary Halstead's mother had been an Onway and consequently Mary's genes were half Onway ones: Mary Halstead's mother (Jane Onway) was the sister of Richard Onway's great-great-grandfather, though when Elizabeth Barton married Richard neither of them was aware of it; as a consequence if we calculate the proportion of the original John-and-Mary genes present in James Onway, when he sold the farm in 1689 after five generations, we find that they were slightly increased by this marriage of distant relations; in this case James Onway had a $\frac{5}{64}$ content of the original John-and-Mary genes instead of a $\frac{1}{16}$ ($\frac{4}{64}$) content which a completely unrelated descent would have given him. It is these reassembled genes that tend to give a family face, or a county or national pattern to communities.

The peasant lines of descent tended to be much closer than those of the upper classes and more prosperous farmers, because villagers travelled but little and rarely moved from their birth-place. This was in contrast to the custom in Tudor times of sending upper-class children away from home. But if a line of descent is traced back far enough a certain intermarriage between lines must be uncovered, as will be found from a simple calculation.

A man has two parents, four grandparents, eight great-grandparents and so on: a generation may be taken at twenty years so that from the building of Bartons End till today some twenty-one generations have passed, which means that the present owners of the house would be descended from over two million* ancestors had no intermarriage of lines taken place. This would be half of the population of Tudor England. In fact, of course, as a man goes back in his ancestry the lines inter-marry to a considerable extent and he would find himself descended from a much smaller number of Tudor ancestors than two million. Nevertheless, though the house has passed by sale to four different unrelated families, the restorer of Bartons End was more than likely to be carrying some of the genes of the original owner, simply because they were both English; but his son is less likely to have so many of them because his mother was Welsh, and thus tended to bring into the family a rather different assortment of genes than an Englishwoman would have done.

What led the humans to abandon Bartons End for ten years in 1899? It was caused more by chance and social conditions than by any innate biological reason. The technical advances in agriculture and industry which had started in Britain led to great prosperity for an extensive middle class, as well as the aristocracy and, for that matter, for a

* $2^{21} = 2,097,152$.

considerable number of the working class as well, because though nineteenth-century industrial conditions were bad, yet cheap transport, cheap food and a range of industrial goods were coming within the grasp of a class who would never have known them at all a hundred years before. The quietening of the Corsican Ogre in 1815 led a number of army officers to seek farms: the growing industries led many workers to demand more beer for which the Weald farmers supplied hops, to their own considerable advantage: but this very prosperity led many to revolt against the ugliness of the industrial age.

Major Hackshaw purchased the farm in 1818, it passed to his son William in 1833 and both of them prospered, particularly from the barley and the hops. This prosperity meant that the Major's grandson, Alfred Hackshaw (b. 1821) could afford to indulge the revulsion he felt at the ugliness of the age and the coarseness of farming—in fact he could be a gentleman of private means, free to become an aesthete, to live for the arts and add embellishment to life. In 1852 Alfred inherited the farm from his father—it will be remembered that he was living in Venice at the time—and let it to a tenant. Alfred's brother David was in the army, and his son Oscar was sufficiently prosperous and well educated to get into the Indian Civil Service: when Oscar inherited the farm he decided to sell the land and to keep the house for his retirement: the prospect was coloured by a memory of a sunny June day at Bartons End, contrasted with the cruelly blazing heat of his bungalow at Allahabad and the fact that when the letter arrived telling him of his legacy he was having an acrimonious discussion with his tiresome khitmagar. On second thoughts he decided to get rid of it and he was advised to sell the farm and house separately. The farm sold easily, but nobody seemed to want the house.

In 1556, after the humans had settled in the house, the animal population grew rapidly.

As winter's approach chilled and bared the ground the field-mice crept under the doors and crouched, bead-eyes staring, whiskers trembling, ready to scurry back at the first creak. At last in the sleeping darkness, daring to dart forward into warm corners, they found crumbs and grains of barley to nibble. This was a good place to live.

'I saw signs of mice,' said Mary Barton. 'We must have a cat.' And so when the cat belonging to her father-in-law next had kittens she asked for one and the tiny creature came to live at Bartons End; but before it had grown old enough for serious hunting two families of mice had been reared in the nest behind the wood pile.

Needless to say cats proliferated at Bartons End, and though they shared the premises with the domestic dogs they lived a life of a very different character, for the dog depends on man and in the main is subservient to him; the cat on the other hand does not appear to be beholden to man at all. One must be careful of attributing to these animals emotions and thought-processes which are human: nevertheless, it is of interest to express the situation in these anthropomorphic terms and then see how it can be explained in other terms. In return for its food and lodging the dog appears to worship man and be ready and eager, on the whole, to obey him: in fact between no other two kinds of animals is it possible for such firm, lifelong friendships to develop as between man and a dog. Dogs have been known to defend the bodies of their dead masters; to starve rather than take food from strangers and have displayed countless other forms of apparently disinterested activity for the benefit of man: in contrast, the cat does not seem to have any respect for his host and will only obey him if he feels like it. 'I am not a friend and I am not a servant,' says the cat in the *Jungle Book*, for 'He walked by himself, and all places were like to him.'

Let us compare these two animals. In the first place their natural history is very different. Cats are solitary, nocturnal carnivores: dogs are daylight creatures hunting in packs and have thus learnt the advantages of co-operating with others. Both spring from a common ancestor—*Miacis*—which flourished some forty million years ago, in the Eocene geological period. The cats have not changed greatly since then whereas the other branch that split off from *Miacis*—the *Cynodictis*—split again and became bears, dogs and raccoons. A cat retains more innate fierceness than a dog; as it leads a solitary life it has had to develop and use more cunning in order to survive, whereas dogs, giving up an arboreal life, have lost the power to retract their claws and have developed instead a swiftness over the ground in order to run down their prey. Cats have retained the power to project and withdraw their claws under sheath-like pads which enable them to climb trees, and to jump and move with extreme quiet in order to stalk and surprise their victims. Both cats and dogs have flesh-tearing teeth, but dogs have long jaws for snapping at their food, whilst cats have short ones which enable them to seize and hold it.

The cats developed a wide-opening iris to the eye, which permits them to take advantage of dim light at night, but nevertheless it can rapidly be shut to a slit, in order that the eye can also be used in daylight. The cat has enormously powerful muscles for its size, adjusted to a skeleton framework which allows it to make a powerful, silent, killing spring common to the whole family from cat to lion. They walk on their toes, do not run like a dog, but trot or proceed in enormous bounds when they want to cover the ground at speed. This enables them to close with their prey at a great pace in the last dash, which is so important to success in this field. Their senses are acute. They have a particularly sensitive perception of moving objects, for it is a matter of importance in stalking game to be aware of the slightest movement of the quarry. The sight of dogs is not as good as that of cats and they are colour-blind, but they are, like cats, sensitive to movement. In cats the sense of hearing is acute, except in blue-eyed white long-hairs which are always deaf, apparently a case of a linkage of a recessive character (deafness) with this particular eye and coat colour. Their sense of smell is well developed, though not employed for hunting to the same extent as is that of dogs. Cats appear to get pleasure from certain aromas, such as catnip and valerian, and may be said to have an addiction to them.

The wild cat has a wide range of food; it consumes all forms of rodents, birds, fish, frogs, certain insects and spiders, grass and fruit; whereas the natural food of the dog is nearly always carnivorous, even carrion, with a certain amount of vegetable food obtained from the intestines of the herbivores killed. Robins have developed a protection against cats in that they cause sickness when eaten: cats rarely eat a second robin, though some, not all, will continue to catch them. It is sometimes thought strange that cats should be so fond of fish when they dislike water; but this dislike is more apparent than real, for all cats can swim and some do it regularly for pleasure.

Can we now explain the difference in the attitude of cats and dogs towards man? Both relationships can be seen as associations for the mutual advantage of host and guest, which I discuss again later in this chapter and more fully in Chapter Eleven. Of the two associations, man/cat and man/dog, the latter is the older: in giving up his pack and attaching himself to man the dog has become that much more dependent on man's favour. Dogs followed primitive men, cleaned up after their kills and from this soon were being used by men, to the benefit of both groups.

Cats were a later addition to man's circle, probably attracted to it by his fire. They caught some vermin for him but were tolerated more for their elegance and totem value in religious ceremonies (they had a part in the oldest of the Ancient Egyptian religions). They may also have had some value to primitive man as food and for their skins, just as dogs did.

A cat can quite easily revert to the wild state; it often does, though it does not then lead so easy a life. The dog, on the other hand, cannot, for in the first place he must form a pack to be really successful and secondly, whilst man may merely dislike a wild cat's activities, every man's hand is against any dog that attempts a wild life because it will do so much more harm (it will attack his sheep, for example). Worse still the dog attempting to live wild was often thought to be rabid and was much feared. The grown kitten escaping to the woods has a good chance of surviving; the young dog a poor one. However, neither the cat nor the dog sitting by the fire is conscious of this, so the differences in the instinctive pattern of behaviour must arise more from the different life-patterns, aided no doubt by the fact that abandoned cats and some kittens have survived in the wild state and subsequently mated with tame domestic cats, which would strengthen the independent attitude of their descendants, whereas abandoned dogs and puppies have rarely, if ever, lived for long. The dog is more dependent on man than the cat, and Lorenz has suggested that the dog has transferred to man the devotion and loyalty he once showed to the leader of the pack.

The Bartons, when they first took up residence, not only had to have cats in the house, but in the farmyard as well, and soon the problem of numbers arose. The yard cats were always chased out of the house and lived a semi-wild life, being fed only with a little milk (and to begin with there was not much of that during the winter) and the birds and rodents they themselves could catch. The house cats had frequent litters and usually all but one or two of the numerous kittens would be drowned. A powerful selective factor was thus at work here, for the successive owners of the farm or their servants were constantly choosing one or two kittens to be kept out of a litter of some six or more. They tended to choose animals with an agreeable coat pattern or attractive features, whereas the more natural selection on the farm tended to develop the best hunters and the most vigorous and healthy specimens. Two separate races of cats did not arise, however, because the farm and the house cats continually cross-mated with each other, which did much to maintain the vigour of the domestic line, and possibly the elegance of the yard cats.

We are only concerned with those in the house: there were always some and there still are two, one a tom, a descendant of the long line of cats associated with the farm and the other a blue Persian neutered male. They answer to the names of 'Tress' and 'Bee' from their full titles, used on ceremonial occasions only, of 'Trespassers Will' and 'Be Prosecuted'.

Prudence Onway's maid, Hannah, was very fond of cats: she once rescued a young abandoned kitten from a hedgerow; it became much attached to the serving-woman and used to ride on her shoulder at times. Hannah had once chased some boys who were stealing apples out of the garden, and the cat had sprung from a tree to her shoulder. The boys denounced her as a witch in the village, saying the cat was her familiar spirit, and indeed it might have gone badly with the poor woman had it not been for two things. Firstly Prudence Onway roundly abused the parson and the bailiff for their stupidity, saying that if an accusation of witchcraft were to be a defence for apple-stealing there wouldn't be a tree, aye or even a sheep, safe in the country. (Sheep-stealing was a crime that really moved people in that age.) Secondly, Matthew Hopkins, the witch-finder, had recently been condemned and drowned as a wizard himself, after causing the death of a hundred people in East Anglia alone. The general hysteria was giving way to a more careful examination and people had begun to doubt the evidence of such interested parties as witch-finders, who travelled the country with what today would be called a big expense account. Hannah lived to a ripe old age, closed her mistress's eyes at the last and was always grateful for Prudence Onway's spirited defence.

Two dogs had come with the Bartons when they moved into the house—a 'sitting spaniel' and a toy dog, or cur as these were called in those days, when there was not the same fine distinction between the breeds as today. John Barton's dog was a bitch called Smoake, which he used in hunting his game. These dogs were called spaniels because the breed was said to have originated in Spain. They were springers, and their chief use was to point the game, which was then flushed and captured with hawks, or else stalked and a net thrown over it.

Many were the different kinds of dogs that came and went during the life of the house. In the early days there were three recognized types and they are described by John Caius (the founder of Caius College, Cambridge) in his book *De Canibus Britannicis* (1570) as: '(i) gentle— serving the game, (ii) homely—for sundry necessarie uses, and (iii) currish—meete for many toyes.' However, somewhat earlier, Dame Juliana Berners, the sporting abbess of Sopwell, in her *Boke of Haukyng and Huntyng*, published in St Albans in 1476, lists greyhounds, spaniels, mastiffs, terriers and tryndel-tayles or sheepdogs. The Bartons also kept a pair of greyhounds for running down hares, for rabbits were comparatively scarce in those days, and sheepdogs for the management of the flocks; but they rarely came into the house. In Tudor times mastiffs were used as a protection against robbers, to bait bears and to hunt deer and wolves. Wolfhounds, particularly the Irish wolfhound, were also used for this purpose, though the wolf was almost extinct in Britain by the time Bartons End was completed.

The squires and prosperous farmers in the neighbourhood always had a great interest in hunting, and the four hundred years saw a great change in both the animals used and hunted and in the methods of the hunt itself. Although hounds were rarely bred at Bartons End they were frequently raised there in the late seventeenth century, for there were but few of the local farmers who would refuse to 'walk' a puppy when requested by the master of the local hunt; and woe betide them if they did, for it was tantamount to a declaration of war with authority. Occasionally a couple of hounds would be boarded at the farm, and it was a thrilling sight to see them respond avidly to the call of the horn when the huntsman rallied them by this simple means on hunting days. At first deer were the principal animals of the chase and the huntsman was not averse to wounding the noble beast with a bolt from a crossbow and letting the hounds finish it off. Two types of raches or running hounds were used, the southern-hound being a heavy, deep-tongued, steady linehunter, and the northern-hound who was a much lighter and more active creature, from which the modern 'pace' foxhound is descended. Both appear to have been bred originally by crossing the bloodhound with a greyhound. The fox was hunted because it was a nuisance to the farmer, killing lambs and poultry, and the hare for the joy of the chase alone. Otters and badgers were also pursued.

Neither horse nor hound had anything like the pace or endurance we see today, which was why all sorts of what we would today consider unsporting devices were used, such as the crossbow. Greyhounds might be set to course deer or the fox. At court hunting etiquette might be strict, but among the squires any method that got results was welcomed.

The fox was originally hunted on foot; the earths were not stopped and relays of hounds ran the quarry home where he was either trapped or dug out, because at that time hounds, particularly the southern-hounds, were not good enough to sustain a long run; nor was the hunter horse much better. As the farms grew and the forests shrank, game animals became scarcer and the smaller mounted hunts took to chasing the fox and hare as well as the deer; in fact they would pursue the first animal the hounds came across. As the game got scarcer so did the stamina and ability of horse and hound improve, thanks to the immense amount of study and effort put into the subject of breeding them. Fox-hunting became a fashionable sport in the mid-eighteenth century, when the Duke of Beaufort's hounds at Badminton got on to the line of a fox one day when returning from hunting the stag and gave his Grace such a splendid run that he took it up. George Munroyd became master to the local pack in 1786: he paid a great deal of attention to breeding hunters and hounds and was particularly pleased with getting a bitch from the famous Quorn pack.

The dog naturally brought into the house numerous other creatures associated with it, such as fleas, lice, flies and worms (which I discuss in subsequent chapters), but the greatest danger to humans from the dog was rabies—a terrible disease, abolished in this country early in this century by the combined use of Pasteur's vaccine and the quarantine regulations. The hue and cry after a mad dog, or frequently one only suspected of madness, would cause panic throughout the countryside and particularly among parents, as children were easily bitten and poisoned by even a small quantity of the virus in the saliva. In spite of this dogs were always welcome and encouraged: they kept down rats, leaving the mice to the cats, tolerated the household cats and chased off strange ones, and were admirable companions to most of the human inhabitants of the house. At the present moment Sloane Dunchester has a chow bitch with a litter of three new puppies, looking very like a collection of children's teddy-bears, a joy to everyone and not unprofitable to her. This breed has a longer history in the south of England than might be imagined, for Gilbert White described a pair brought from Canton about 1784, where they were bred for food, being fed mostly on rice. They had the same stance, coat, and black tongue and mouth, the tail curled over the back and the bare spot to receive it then as now, but the head was slightly peaked whereas now it is flat. Possibly the specimens White describes were poor ones and not typical of the breed.

By the time Mary Barton lay in bed with her first-born, that short-lived son who died in infancy like nearly half the babies of that century, but for whom Mary grieved as if no baby had died before, the house-martins were nesting under the eaves above her bedroom window. And the maids were dusting away cobwebs spun by spiders from beam to beam, the crickets were singing beneath the fireplace and the woodworms gnawed patiently on.

The mice were the third mammal to enter the house: they are very 'plastic' animals, for they can very readily adapt themselves to varying conditions and then be so altered as even to pass as a different species. Six kinds are met with in England—the house-mouse, the long-tailed field-mouse, the tiny harvest-mouse, the dormouse, the continental dormouse and the yellow-necked mouse, but it is the first, the house-mouse, which is most important in the countryside, and it is the only mouse that established itself at Bartons End. Strangely enough it is the house-mouse which is most seen in the harvest field, not the long-tailed field-mouse, which is frequently suspected of being after the grain.

The ancestors of the house-mouse lived on the open steppes of Russia where they consumed the seeds of plants, so it is not surprising that they attached themselves to man when he became a farmer growing cereals; we describe them as commensals—feeding at the same table as ourselves—and, associating with man, they have now spread all over the world.

Mice

In any country area there are usually two communities of these house-mice—the house-dwellers and the field-dwellers. Though they belong to the same species of animal, these two races do not usually have much to do with each other. There is in fact some basis for the stories of the town mouse and the country mouse, since there exist these two populations, each a stranger to the other's environment. However, some of the house-mice go out into the fields in summer, where they continue to breed and seek to get back to the shelter of their commensal in the autumn. The other community of mice (the same species at present, be it remembered) live and breed wholly out of doors where they construct burrows and nests in hedge bottoms, banks, tree roots and so forth. It is the house-mouse which is really harming the farmer both in his home, his fields, and yards, not because the direct damage the mouse does is dangerous—in the house the quantity of food eaten is not great, though in a stack it can be serious—but because mice spoil more food than they consume and can carry serious parasites and diseases of man and animals. Mice will be attacked by fleas, which in turn can carry and transmit the plague to man. This scourge came to Bartons End in 1665, and killed Mark Onway, who had been bitten by an infected flea on a visit to London, but it did not spread, perhaps because Prudence, Mark's wife, who was a very house-proud woman of Puritan stock, would not tolerate mice or fleas in her household and was always beating the beds and scrubbing and polishing the floors so that it was by a great mischance that Mark happened to be bitten by an infected insect.

Another danger from mice, which persists to this day, is trichinosis, a most distressing malady caused by a roundworm which passes from host to host when certain raw or insufficiently cooked flesh is eaten. Mice are not eaten by man (except the edible dormouse) either raw or cooked but pigs sometimes kill and eat mice, and dead mice and rats were frequently thrown to pigs—perhaps they still are at times—and these animals become infected and then infect man in their turn. Pork, ham and bacon have to be very thoroughly cooked to destroy the encysted trichinosis worms in their flesh.

The original habitat of the mouse was the open plains and its numbers under natural conditions were controlled by predators, such as weasels, stoats, hawks, owls, cats and so forth. In the wild state mice lived in burrows and thickets, where they were reasonably safe from attack, and came out in the open to feed, when they were much exposed to their predators. By contrast, in the house the mice may be under cover the whole time, or only exposed to one predator—the cat—for very short periods, so it is not surprising that the animal has adopted a very advantageous commensalism. It is no wonder that man has had to reply with traps and poison as well as cats.

Sometimes plagues of mice and voles arise, due to some change, possibly quite a small one, in the biological balance of a neighbourhood: they happen at times when there is a lot of waste land, for here, under the right conditions, a big population can build up, usually due to the abundance of some particular seed plant. Waves of mice and voles then depart from these lands and invade the fields and houses of men: their numbers are very impressive and spectacular but probably these plagues do not cause as much harm to the farmer as the steady annual depredations of an ordinary mouse population in ricks, stores and houses. The plagues of mice disappear seemingly as quickly as they have arisen and their vanishing does not seem to be due to any particular disease, parasite or predator, though naturally all can be found during a mouse or vole plague. For instance, the short-eared owl is much drawn to any such area but even so can have but little effect on the vast numbers present during a plague. Many mice are found infected with tuberculosis, but this again does not seem to be the reason for the reduction in numbers, as the disease does not shorten their lives to any appreciable extent.

There is a protozoan parasite, *Toxoplasma*, which attacks the brain of those small mammals and while such a disease could decimate a population it has not actually ever been found doing so.

The population builds up because of favourable circumstances, not the least of which is a good supply of food. The young mice are sexually mature at three months old; the females can breed over a long season and carry their young for a period of surprising variation—from twelve to twenty-one days—and it would seem that an abundant food supply leads to a shortening of the period of gestation. Very soon after giving birth to a litter they come into heat again so that soon another litter is on the way, to consist of some five or six blind, naked pink young creatures who are ready to leave their mothers at less than three weeks of age. Obviously all the individuals of a fast-breeding species such as the mouse cannot survive or the world would soon be knee-deep in the creatures: for a population to be stable a certain 'survival rate' must prevail and it can be shown mathematically what this rate is.* To maintain a stable population—for instance a 'normal' one, say fifty pairs per acre—if the number of litters per three months is two and the number of mice per litter is three, the survival rate of mice must be sixteen per cent.

Let us suppose a change occurs and the number of litters per quarter rises to four and the number of mice per litter to the same number, which is not a very big change; in these new circumstances the survival rate must be reduced to just over one per cent in order to maintain population equilibrium. This is a big difference from a rate of sixteen per cent and when the change first occurs the rate is not likely to drop much, if at all; as a consequence the population numbers rocket upwards and a plague occurs. The competition between individuals then reduces the survival rate; if this falls below one per cent the population falls again; if, as is most likely, the mice per litter and the number of litters per three months also fall then the collapse of the population is as rapid as its increase. It is comparatively small changes in the rates of breeding and survival that make these big differences in numbers. The same thing can be seen in the rise and fall of locust numbers and is based on the fundamental mathematical fact that a geometrical progression rises or falls in its later terms by big numbers. The mouse's life span is a year or eighteen months and the sudden decline in population is accelerated by there being a considerable number of middle-aged individuals in the plague community who have a short life expectation, as well as a lower rate of breeding.

The clearing of the Wealden forests led to a big aftergrowth of bluebells and grasses, which much encouraged the multiplication of wild mice and caused some plagues at Bartons End in the early days, but as soon as all the countryside was settled, about 1650, such plagues were known no more.

---

* The survival rate to maintain stability, can be expressed by the formula:

$$S = \frac{1}{(1 + \frac{r}{2})^n}$$

where r = number of mice per litter and n = the number of litters per 3 months.

The house-mouse is able to tolerate a wide range of temperatures provided it has food and shelter, though in general small mammals are less able to resist low temperature than big ones. The onset of the cold drives mice back to the house but observations in recent years have shown that mice in refrigerated meat stores have remained perfectly healthy under conditions of seventeen degrees of frost. Under natural conditions mice would not survive at this temperature as they would not have an unlimited supply of food, but these lived mostly on liver, which gave them plenty of protein and vitamins. E. M. O. Laurie found that they ate about one-third more food than did mice under more normal conditions, as is only to be expected. Mice need protection against excessive heat and long exposure to sunlight.

They do not need very much water—but though they can survive for a long time without it, they will not increase in weight or generally thrive unless they have some, or at least moist food. In hot weather in such a place as a wheat rick, lack of water may be a serious handicap to mice and they will leave the place to seek it, but in normal weather they can get what little water they need by burrowing to the surface thatch or the windward side of the stack, where there will be water after rain or heavy dew.

When food is short, or lacks some necessary element such as a vitamin, the young in the nests frequently have their heads bitten off by the mothers, which suggests that this is a mechanism for adapting the size of the population to the available food supply. At Bartons End the wheat stacks supported a far higher mouse population than the stacks of barley and oats, for wheat is a more satisfying food than the latter cereals. A diet of one cereal only is not ideal for a mouse, but mice increase in a stack because they are sheltered from their enemies and from extremes of climate. Here the numbers rise up to a certain point of infestation but do not pass it. As the stack ages it often collapses, due to the activities of the mice.

When the mice first entered Bartons End they took about three weeks to become settled, during which time they made a very thorough investigation of the place and kept their feeding to a minimum. They rely very much on their knowledge of the neighbourhood and do not like it when furniture or store rooms are rearranged; in fact one of the ways of keeping down the depredations of these rodents is continually to rearrange their environment. The danger period for the mouse is when he moves from cover to food; he likes to be well acquainted with the routes and to use them at night when predators will not see him so well: but the mouse is a plastic animal and if under any particular conditions the daytime is the safe period he will adapt himself to searching and feeding in the light. The kitchen of the night-watchman tends to be explored during the day.

Mice do not have very clear sight, nor are their eyes particularly well adapted to nocturnal life. Though they do not see shapes well they are very sensitive to the intensity of light. They are very quick to notice movement and either this or a change in light intensity immediately produces their attentive reaction, when the animal becomes alert and ready to run away if danger threatens. They react more at night and obviously prefer to be active then rather than during the day.

Their hearing is acute, especially for high notes: the squeaks in which they indulge are a language of a sort in that this noise conveys information about danger or food supplies and seems most useful to them. They are frightened of loud noises and instantly seek safety in flight when one occurs.

Their sense of taste and smell is well developed: smell is of great assistance to them in finding food and they remember tastes sufficiently well to avoid for a time at least any food which has disagreed with them; this means that if a mouse has been made ill by a small quantity of poison bait it will avoid that bait for several weeks, until it has forgotten the experience.

The awareness or perception of their own movement—the kinaes-thetic sense—is very important in mice: they rely on it for finding their way in the dark and to escape from danger. A mouse likes to explore its neighbourhood very thoroughly and along all the various routes it learns the necessary sequence of muscular movements: it is as though it said to itself 'From the hole in the skirting board, twenty-eight paces to the door, turn right under the door, thirteen paces forward round the coal scuttle, twelve paces half left, twelve up to the chair, then jump to the table and food (the remains of the humans' supper) will be found.' The mouse, of course, has no such actual thought in its head but it acts as if it had; it goes through this sequence of remembered muscular movements to produce the effect. This is a reason why the mouse at any one time ranges only over a strictly limited area: if it reaches a new zone it has to learn this new field before it can feel at home.

When a new object appears in their particular environment the mice do not avoid it to the same extent as do rats but rather explore it carefully, so that the experience can be added to their kinaesthetic memory. This sense is not absent in man, but is not normally used by him, because other entities, mostly sight and memory, perform the function better and more easily. However, when a man loses his sight he greatly develops his kinaesthetic sense and uses it continually to find his way about, a fact effectively illustrated by Bulwer-Lytton in his novel, *The Last Days of Pompeii*, when Nydia, the blind slave-girl, leads Glaucus and Ione to safety through the murk of the ash-inundated Pompeii: 'Her blindness rendered the scene familiar to her alone.'

Mice have a keen sense of touch which enables them to appreciate small, narrow places where they can take cover when danger threatens. It is of great advantage to a mouse to have a small range, as this reduces the chances of its being attacked: they only wander far and wide when food is short and at other times will keep their area of activity as low as is consistent with getting an adequate supply of food.

Mice must continually wear down their teeth by gnawing; if they are unable to do this the incisors will become too long and they can no longer feed.

During the first hundred years of the house's existence the mouse population at Bartons End was nearly always high because the corn ricks and the barns provided a constant reservoir of population from which the depredation of cats and traps could continually be made good. In 1655 the rick yard was equipped with rat- and mouse-proof platforms, supported by the stone mushrooms which prevented the animals getting into the rick, or getting back into it once they had left it, as usually a few mice are built into a corn rick as the sheaves are piled in. This greatly reduced the population of the house.

A few mice in the house did comparatively little harm and at first the servants had rather welcomed them as pets, but as they became numerous Mary could not stand the sight of their droppings and the depredations they made on the food: soon the cat and its descendants had considerably reduced the nuisance, though the house was never entirely free from mice again until the summers between 1899 and 1909 when it was uninhabited by man. As there was no food being dropped or left unprotected, there was no reason for the mice to remain in the place during the summer, though a few did come back to shelter in the winter. They did not thrive there in the absence of humans. For instance, when man left the island of St Kilda in 1930 the house-mice soon became extinct there—they could not live without their commensal.

The singing mouse which old Hannah (Prudence Onway's maid) had as a pet in 1672 was really a mouse suffering from bronchitis: it was a dangerous pet to have in those days, not because the bronchitis was catching, but because of the risk of the creature carrying plague-fleas.

In 1845 a small strip window at the top of the south gable of Bartons End was blown in by a gale: it fell on the loft floor and was not noticed by the humans, but the gap it left attracted the attention of some noctule bats who were looking for a winter hibernation site. They soon established a considerable colony there, which was only briefly disturbed by the clearing of the house on William Hackshaw's death in 1852. During the tenancies of Emmanuel Burrows and William Daker the window was not repaired and the noctules found the loft most useful as a shelter. They also found a source of food in the shape of adult wood-beetles, as well as the other insects and spiders. It will be remembered that the roof fell in at the north end in 1903, when some pipistrelle bats immediately took advantage of the opportunity and also established a colony, this time at the north end of the house.

Bats are the only mammals which have adapted themselves for true flight. The arms, hands, with their fingers, the legs and tail form a framework on which is stretched the membrane making the wings, and have been much modified in the process. The forearm is extended and the bones of the fingers, except the first, are elongated, serving to spread the wing, which is just two layers of fleshless skin. The leg is also used for the same purpose, and it likewise has been modified and in effect turned round, so that the foot faces backwards. The foot bones are much reduced, appearing as the two little hands by which the majority of bats hang when they are at rest. The powerful muscles of the chest and upper arm are able to flap the wings and carry the creature aloft, where it shows great dexterity in carrying out swift flight, extremely quick turns or hovering in the air as occasion may warrant. The bat is an example of a marvellous adaptation to a special way of life, the principal features of which are hibernation, an insect diet, low birth-rate and the truly amazing powers of flight in the dusk or dark, flights in which food is found and obstacles are avoided with great ease. The smallest British bat is the pipistrelle—indeed it is our smallest mammal and has a wing span of eight inches—whereas the noctule reaches fourteen and a half inches across. ·

The jaws of bats move slightly from side to side as well as up and down, so that they have a rotary action rather like that of a human feeding, which enables them to cut up the hard shells of insects. Only one offspring is born at a time. It has milk teeth at birth, with which it grips its mother's fur, as for a time it accompanies her in flight; when it loses its milk teeth it holds on to its mother by gripping the false nipples in the region of the groin: it is suckled from two true nipples on the breast, mostly from the right-hand one as this is usually larger than the other. When the young bat is ten to fourteen days old it is too heavy to accompany the mother and is left behind in the roost, where it seems to hang quite happily and is fed by the mother on her return from the hunting expeditions. Sometimes the mothers seem to feed the first young bat they come across and at other times to seek out their own particular baby. The young are born in early summer and will be on the wing by August. Soon they are able to hunt for themselves and build up reserves in their bodies to nourish them during their hibernation.

Bats appear peculiarly furry because, though the coat is long and silky, with a downy undercoat, it tends to stand out straight from the body which gives the animal a bristly appearance. British bats are dull in colour, nor is there any difference in colour between the sexes. They have large ears and exceptionally keen hearing, which enables them to receive ultrasonic sound waves. Their eyes are small and not particularly adapted to nocturnal, or twilight, flight. 'As blind as a bat' is a common phrase. On their nose they have a 'noseleaf' which seems to be an organ enabling bats to sense objects at a distance.

The bats' powers of flight vary greatly between species; some may describe a pattern which is very vigorous, swift, high and full of sudden turns, whereas others flutter about near ground level. Many bats adapt themselves to a regular flight path at a regular time, which of course

Noctule bat

coincides with the presence of their particular insect food: they are fond of the edges of woods, or a glade through them, or of flying over water where there are aquatic insects, or through a farmyard. They like to break into flight from a height, but they can jump into the air from the ground if it is necessary, by making a combined effort with legs and arms. They tend to alight head upwards from their flights and then turn round and hang downwards by their feet, perhaps holding on as well by the wing spurs, which are really their thumbs. The horseshoe bats have a most peculiar method of alighting from a journey. They always choose a ledge, beam or projection from which they can hang freely, so they use caves, cellars, or roofs: as they approach the spot they turn a forward somersault in the air, close their wings at the same time and grip the perch with their tiny feet. It is very seldom that this complicated manoeuvre is unsuccessful and, if it is, the bat falls, is soon in flight again and lands on the perch at his second somersault.

Bats can skim very close to the surface of water and if they fall in through a miscalculation they do not exactly swim, but by spreading their wings and flapping, they propel themselves towards the bank and usually escape.

The most remarkable thing about bats is their agility in flight: when in the air they can avoid obstacles and obtain their prey with equal facility. This is done by a form of echo-sounding which has been compared to radar, but though it is similar it is by no means the same thing, because radar uses electrical impulses and the bat sound impulses, albeit ultrasonic ones. Electrical impulses move with the speed of light, whereas the bats' signals go at the speed of sound, nearly a million times slower. In the eighteenth century the zoologist Spallanzani discovered in Italy that bats did not find their way about by sight and in 1920 Professor Hartridge at Cambridge suggested that their remarkable powers might be due to the reflection of sound from near-by objects, which theory was proved by Griffin and Galambos in the USA in 1941.

A bat's hearing is far more sensitive and is extended over a much wider range than our own, as, of course, is that of many other animals as well. We can hear from 16 vibrations a second to about 40,000 per second when we are young, though it is said that no one over forty can hear a bat: for these creatures make their ultrasonic signals at pitches of from 30,000 to 70,000 vibrations per second. They send out a series of signals at rates of about ten squeaks per second when at rest, to thirty per second in flight, which are reflected back from objects in the bat's path, are received in his capacious ears and tell him a great deal about the nature of the space in front of him. The more a bat wants to know about the objects ahead the quicker will he send out these signals. If you shout across a valley to a rock face you may get an an echo back and the time it takes to come back, its strength and clarity can convey information about the distance and nature of the rock on the other side; if it comes back quickly and clearly it must be quite near and be relatively smooth: if the echo is ragged it must be a broken face ahead, if it takes a long while it is far away, and so on: were we to practise listening to these echoes with our eyes covered we no doubt could learn to gather a lot of information about the reflecting object in our path, and that is just what the bat has done.

From the ultrasonic echoes he can tell how far away a thing is, its size, something of its nature and the way it is moving, which enables him either to make a swift turn to capture it or to avoid it, just as he pleases. If you shout at your rock face and then shout again before the echo has come back you will most likely drown the noise of the echo with your second shout: a group of children are often disappointed with echoes because they cannot contain their excitement and little Jennifer's shout drowns Roger's echo, which in its turn is spoilt by baby Pamela's belated attempt.*

The same sort of thing could happen with the bat; he could get the wrong sort of information back from his echo or even no intelligence at all if he lets the outgoing and incoming signals confuse each other; this means he must get back the echo to his first signal before he sends the next and moreover he must not deafen a sensitive ear, poised waiting for a faint echo, with the loud signal being sent out. If we shout at a rock face quite near us we do not hear the echo because we have made our ear relatively insensitive with the shout. The bat avoids this trouble with a special muscular mechanism which puts the ear out of action as the signal is sent out and then immediately releases it, ready and fully sensitive for receiving the echo. It is a truly wonderful and amazingly successful system though not unknown to man: for instance, if during

* This was a dramatic device of the Elizabethan playwrights, where the echo would give back a phrase with a far more sinister meaning than the message sent out, because the first

the ringing of a bell you yawn heavily you will put your ears out of action and not hear it.

It seems strange that many bats flying together do not confuse each other with their ultrasonic signals and echoes, but the explanation seems to be that each bat has a distinctive signal of its own: it is able to recognize its own voice. Moreover it is likely that these signals die away quickly as their effective range only seems to be about five yards. Bats can turn very quickly in their flight; they must be able to do this because the danger signal may only be received when they are very close to an obstacle. They recognize desirable insects on the wing, that a twig ahead has a spider on it and so on, and rarely make mistakes. A bat will occasionally dive at a fisherman's fly as it whirls on his cast: bats have been caught this way but nearly always in the wing, not in the mouth which suggests that the echoes being received told the creature at the last moment that the object was not an insect, but that his turn away was not fast enough to avoid the hook completely.

These ultrasonic noises require considerable effort on the part of bats and the larynx in these animals is a very sturdy structure operated by powerful muscles: anyone who plays a wind instrument knows that you must put more energy into getting the high notes than in playing the lower ones, and still more must be used to get those higher still, the

words of the echo were drowned by the last words of the message.

ACT V. SCENE III: Milan, part of the fortifications. Enter, Antonio and Delio. There is an Eccho from the Dutchesse (of Malfi's) grave:

ANTONIO	. . . But all things have their end:
	Churches and Citties (which have diseases like to men)
	Must have like death that we have.
ECCHO	*Like death that we have.*
DELIO	Now the Eccho hath caught you.
ANTONIO	It groaned (me thought) and gave
	A very deadly Accent?
ECCHO	*Deadly Accent*
ANTONIO	Eccho, I will not talke with thee
	For thou art a dead Thing.
ECCHO	*Thou art a dead Thing*
ANTONIO	My Dutchesse is asleepe now,
	And her little-ones, I hope sweetly:
	Oh Heaven
	Shall I never see her more?
ECCHO	*Never see her more.*

ultrasonic vibrations, which we cannot hear at all. Bats, however, do have two audible voices as well; one is the high-pitched squeak which we sometimes notice, particularly when we are young, and the other is a low buzz. The first is a language of sorts, somewhat akin to the squeaks of mice, and serves, in a primitive way, to communicate information among the colony; the other is a product of the ultrasonic emanation. These ultrasonic signals are given out in short bursts and a series of these bursts will itself make this low audible buzzing note: when it is heard it means that the bat is sending out its echo-signals.

Bats seem to have another and as yet unexplained sense, giving an awareness of objects near them: it is not the echo-sounding effect because it happens even when the bat is torpid and asleep and is thus not sending out ultrasonic noises: it has been described as feeling at a distance. A sleeping horseshoe bat will become aware of an object coming near it and show this by drawing itself up towards its perch, even though it be hibernating and too torpid to fly. It is possibly some reaction to the approach of enemies, to try and make itself smaller and less conspicuous and to get ready to drop or spring away. It is when the bat is enveloped in its wings that it seems to be able to sense a change in the environment: a possible explanation of this strange reaction may be that the wings are very sensitive to any sudden, however slight, change of temperature caused by a warm object—for instance a predatory cat—or a colder one, such as the head of a stick being advanced to push it down.

Mating of bats takes place in the autumn, but the sperm is stored in the female during winter hibernation, though mating may also take place again in the spring. The actual ovulation does not occur until the spring when the egg is fertilized by the stored sperm. Apart from these periods most bats roost in groups with the males separated from the females. The females are usually two years old before they breed and the period of gestation is about six weeks. The young, one at each birth, are born in the early summer. They are blind, pale and naked and cling to their mothers' fur as was mentioned above.

The food of bats is almost exclusively insects and spiders, though Gilbert White accused them of attacking bacon hanging in the chimneys for smoking: in this case they were more likely feeding on insects living on the bacon. They nearly always catch their food when flying and if the insect is too big to eat at once the bat rapidly bites off the wings, or useless appendages, and tucks the juicy parts into a pouch formed by the tail being bent forward—the interfemoral pouch. They can pick up water from the surface of ponds when on the wing.

In spite of their reputation bats are very clean creatures and as soon as they alight and are hanging upside-down start an elaborate toilet, comb their fur with their feet and clean their feet with their incisor teeth. Nevertheless they do suffer from parasites, such as a peculiar large wingless fly, *Nycteribiidae* which sucks their blood. These flies are tolerated by the bats, who make no attempt to kill them; they just emerge from the fur and run quickly to another spot as the bat combs them out. It is as if, as L. Harrison Matthews points out, we were to tolerate several large crabs hidden in our clothing.

Bats spend the winter in hibernation: as they live almost exclusively on insects taken on the wing there would be very little winter food available were they not to do so—they must either lie torpid, or migrate. Sometimes bats have different summer and winter quarters and sometimes they use the same place all the year round. The natural winter habitats of bats are caves and hollow trees, which have a distinctively limited distribution in nature so that the advent of man with his easily penetrated buildings, above all on the farm, has been greatly to the advantage of bats, particularly when by his activities man encourages insect life. Bats will form colonies in cellars, lofts, barns, church roofs and towers, in fact any place where there is shelter and they can be reasonably quiet. They do not mind very damp places; seemingly they enjoy them, for in some localities they choose the humidity is so high that the water forms in beads on the bats' fur.

They prepare for hibernation by feeding enormously and accumulate a thick layer of fat beneath their skin. They then get together in a mass in their chosen retreat, hang downwards from a suitable perch, much reducing their metabolism and becoming quite torpid. The heart-beat and rate of breathing are lower as is their general temperature also, so that their oxygen consumption is only at one-hundredth of the rate they maintain when awake and active. The hibernating bat has curious spells of breathing and then stopping: it will breathe at the rate of thirty intakes a minute for about three minutes, stop for about the same length of time or even up to eight minutes, and then start again: this compares with a rate of two hundred breaths a minute, with no pauses, when the animal is active. During hibernation much of the blood is stored in the spleen, so that it is not circulated so actively as during normal life.

This greatly reduced activity means that the food reserves are only drawn upon slowly, and in fact the hibernating bat is only using food at about one-twelfth of the rate of consumption during activity: even during the summer bats are able to pass into a semi-torpid state, usually, of course, during the daytime, and they are thus able to reduce their food demands, for when they are active they have big appetites and if they were so all the time they would need to eat their own weight of food daily. In the summer colonies during the daytime, not all the bats are in the semi-torpid condition: this would seem to be a safety provision as the merely sleeping bats, who usually exceed the torpid ones in number, can warn and wake their torpid companions should danger threaten. However, it takes a long time to waken a completely torpid animal, from half an hour to an hour: its temperature must rise from about 48°F to the normal of 98°F and its respiration reach the normal of two hundred breaths a minute: the blood must be withdrawn from the spleen and circulated in the usual way by the heart.

In the same way as during the summer bats may go into semi-hibernation, so during the winter they may come out of it for short periods, when they fly about their quarters or even emerge into the open air if the evening is mild: if any food such as hibernating insects or spiders is found, these hibernation-breakers will take it. No doubt this revitalizes them and allows them to resume their winter sleep again without harm.

The facility of going in and out of hibernation means a great deal to a bat's food economy: though protected by long fur the creatures have a big surface area to weight ratio, which means that the heat loss of a bat is big compared with a larger mammal, such as a cow or man; this heat loss must be made good by consumption of food. Moreover a bat uses a great deal of energy in flying, which again must be provided by food; the food in its turn is rather specialized, mostly insects and mostly those found on the wing; which again means that the bat has to fly a great deal in order to find it. Being in the torpid state means economy in the use of food or food reserves in the body. If you consider a sufficient number of bats, say 1,120, to equal the weight of a 10-stone man, the former, when active, would be eating 10 stone of food a day whereas the latter, on a diet of 3,200 calories plus 75 grammes of protein, would consume only about 4 pounds, so we at once see the advantage of being our size and also of not having to fly to collect our food. Had we to do so by our own effort, our muscles and wings would be enormous and we should need a daily food intake of several times our own weight instead of a fraction of it.

The evolutionary process has determined that certain average weights for animals are the best resultant of a number of forces. The smaller the ratio of weight to surface area the fewer calories per kilo will that creature need to keep it warm, but as size increases so must the size and cross-section of an animal's bones increase to support that weight, and this means bigger muscles, which again demand more fuel to drive the limbs. So we compromise between small weight, and thus easily moved limbs with a low fuel consumption, and large size with a low heat loss per unit of area.

Man needs 40 to 50 calories per kilo of body weight per day and the active bat some 700 calories per kilo, but by introducing torpidity into its life-cycle the bat brings this demand down to about 58 calories per kilo per day, which is much more like the human figure. Of course the torpid animal is at the mercy of its enemies, but the bat hanging from the roof of a cave, cellar or a man's house has chosen a spot where these are very scarce.

If we take a much larger animal, such as a cow, we find it only needs about 16 calories per kilo of body weight per day in order to maintain it, which gives us an interesting series, 700 for the active bat: 58 for the torpid one: 50 for man and 16 for the cow.

Bats seem to be induced to go into hibernation more by the high feeding they have indulged in during the summer insect flush than by any other factor, followed by a check to feeding as the evening temperature drops with the onset of autumn, because bats will not come out of their roost for their regular hunting flight if the temperature is below 40°F, or if it is windy.

Another curious characteristic of bats is their low reproductive rate, which begins to be comparable with that of man and is another facet of their very specialized way of life: it has already been mentioned that there is only one offspring at birth. If the numbers in a species are not to

Pipistrelle bat

decline the net reproductive rate must be unity, which means that every female must be succeeded by one surviving female of breeding age. The number of the two sexes born is approximately equal. The females do not breed until they are two years old and only breed once a year, so to ensure a net reproductive rate of over unity a female must live to an age of over four and a half years to have the necessary three offspring. It is very likely that most bats have a life span of some seven years. This is a long time compared with the life of a mouse—a mammal of comparable size—which has a year or eighteen months. Bats undoubtedly live to this ripe old age because they spend so much of their time in hibernation, a state of torpor, or asleep.

Noctule bats flew around the farmyard at Bartons End from very early days and hibernated in the high roof of the barn and the eaves of the stables. They did not penetrate the house itself until 1845. They do not seem to mind rain and will go out hunting on the wettest evening, but they do not emerge if it is either cold or windy unless a succession of such days forces them to face the weather and seek food. They are very variable as to the time and duration of their evening flights: they may come out of their roost, giving their high-pitched *wyee, wyee* squeak in quick succession or at considerable intervals, from half an hour before sunset to half an hour after it and may stay out from half an hour to an hour and a half. There is also a morning flight which takes place about an hour and a half before sunrise and lasts for forty minutes. Each colony of noctules seems to maintain its own hunting zone and to drive other bats away from it.

They fly very high and fast and tend to take long straight sweeps; they do come down if their prey is low and they both eat and drink flying. They cover long distances, and from Bartons End they may even have crossed the Channel for the summer and returned to their cosy attic to hibernate.

The pipistrelle bats moved into the north end of the attic at Bartons End when the roof fell in. They are very small creatures with a wing span of about eight inches. They hibernate in large colonies and often with other kinds of bats. During the summer they may roost in many places, frequently being in different ones by day and night. Like the noctule they do not mind rain but though not so sensitive to cold they too do not like wind. Their evening flight is earlier than that of the noctules and they make many short journeys from the roost usually all through the night: they tend to adhere to a particular route and they had just such a regular track round the farmyard and house at Bartons End. They stop flying at sunrise. They have a flight pattern very different from that of the noctule: the pipistrelle's is one constantly turning, very fast and seldom much above twenty feet in height.

Although the first colony of bats only formed in the house when there was an easy means of access in the mid-nineteenth century, there had been bats present before that. Little Prudence Onway, born in 1654—the sister of James Onway—had kept some as pets; she was from an early age thought to be a strange child. When she was eight years old she could read and write well and by fourteen was quite at home in Latin and French and eagerly read the new poem *Paradise Lost*, given her by her indulgent grandmother, old Elizabeth Onway—the last of the Bartons. Prudence's greatest interest, however, was in animals and plants. She loved Topsell's great picture book *The History of Four-footed Beastes*, which had been a prized possession of her father's—he had died in 1665—and she regretted that so many of its wonders could never come her way. In the summer of 1668 she found a noctule bat in its daytime sleep hanging on a door in the dining-room and determined to keep and tame the creature. She carried it upstairs to one of the empty bedrooms and in two weeks she had tamed it, before her mother or grandmother knew anything about it. Prudence and her brother fed 'Master Milton' as they irreverently called the creature, with beetles, scraps of meat, meal worms, grubs and spiders and soon she was able to astonish the family by opening the parlour door and calling Master Milton to come and sit on her shoulder. In a less enlightened age, said her grandmother, it would have been called witchcraft. James and Prudence soon added two more bats to their collection and after a while there was no need to keep the creatures shut in: they flew round the house snapping up insects, waited for their food and rarely went out on hunting expeditions, as they got too fat and lazy. This was about the same time that Hannah had the singing mouse as a pet. Prudence married a neighbouring farmer's son in 1674: the tame bats were no longer fed so regularly at Bartons End, so henceforth bats were only occasional visitors to the place, being tempted in on exploratory expeditions when looking for roosts, or insect food.

Bats are very old creatures: fossil remains found in the Eocene period—forty million years ago—show they were fully developed by then and the long force of evolution has placed this strange pattern of life on them. The British bats are insectivorous but overseas, particularly in the tropics, they feed on other foods: there are the notorious vampire bats of Central and South America who suck the blood of mammals (they weaken but rarely kill their hosts) and the flying foxes of Australia who eat fruit. There are even fish-eating bats. The British bats are highly specialized feeders and take advantage of one particular supply of food—the insects. The numbers they destroy are enormous and, while many beneficial insects as well as pests must be eaten, there is no doubt that bats were friends of all who lived at Bartons End, because they kept down the attackers of the farm, the garden and the house.

Chapter 7

Spiders

JOHN BARTON's house was built on the far boundary of Richard Barton's farm, which was why it was called Bartons End. The site was a meadow below the wood, which was being cleared very rapidly, mostly for the purpose of making charcoal to supply the Weald iron smelters, and though the first animals to live in the house were the wood-borers, as they were in some of the timber used for beams and floors, the most numerous animals actually on the site itself when the first turf was cut were the spiders.

Spiders were particularly plentiful in the grass from spring to autumn, and as the house rose during the year 1555 a considerable change took place in their relative numbers, as was only natural. The grass-living hunters and sheet-web builders moved away or died and the house-loving web builders colonized the structure as it rose, offering them more and more splendid sites for webs, snares and shelter, till finally the house-spiders occupied the cellar and thatch in large numbers.

Spiders are not insects: they belong to the family of jointed-limbed animals, or Arthropods which also includes such beings as crustaceans, scorpions, and insects. Looking at spiders one sees that they are rather different in appearance from insects: in the first place their bodies are divided into two main portions (as against three for the insects) having the head and thorax united into one piece, and behind this an abdomen: secondly the spiders have eight legs against the insects' six. They moreover do not have a complete metamorphosis as do most insects, who pass from egg to larva to cocoon, pupa and adult, but hatch from the eggs as tiny spiders and grow larger by successive moults, forming a new skin beneath the old one, splitting the old skin off, expanding the new skin and letting it harden. Spiders and insects are similar in that they have no internal skeleton, but a comparatively hard external one, to which the muscles and other organs are attached. Like insects they absorb their oxygen by the diffusion of air through tiny vents on the surface of their bodies. Spiders have either two, six or eight eyes, and as these animals inhabit an enormous range of different sites, from the bright, sunlit fields to the deepest sunless caves, their powers of vision vary enormously, as indeed do all their other powers, for their 'plasticity', we might say adaptability, has enabled species to develop and specialize in colonizing differing areas, sites, or kinds of food. Some spiders will hunt their victims by day, others by night, some will build webs to catch a certain type of insect, some tiny spiders will live in other spiders' webs, feeding on the crumbs which fall from the rich man's table. All this has been brought about by the power of natural selection and enables them to prey on the whole field of insect life.

Spiders have powerful jaws with which to seize their prey and in most cases the last of the fangs is hollow with a minute hole at the tip, through which they inject the acid, colourless, liquid poison with which they paralyse and kill their victim: this has been secreted by the poison glands within the body. They also have a pair of palps, which serve somewhat as hands, being useful in feeding and other activities.

The best known characteristic of spiders is their power of spinning silk from which they build webs, but it is not a characteristic of all of them, because some live by hunting their food and some by spitting sticky gum at insects, thus pinning them down. The silk is produced from special organs at the rear of the abdomen; and each may have more than a hundred tiny openings (spinnerets) through which the silk is projected. The silk glands secrete a fluid which is squeezed out through the spinnerets in the organ and the filaments harden on contact with the air. Its relative strength is enormous. The spider can move the organs, which project like tiny short fingers, so that if the tips are separated a wide band of silk can be formed, or if kept close together a strong compact thread results. Some spiders have a comb, called the calamistrum, on the hind leg which helps form the sheet silk and also a special central silk-forming organ with very fine openings, so that sheet webs, which so collect dust and fragments in our lofts and old buildings, are easily made. Two kinds of silk may be spun, the first with sticky beads on it on which insects may be caught, and the other plain, strong inelastic threads.

In addition to making webs the silk is also used for egg cases, for making tubes or tents in which to live and as a means of locomotion. On a warm day with a gentle breeze and rising air currents the spiders, particularly the young ones, climb to the top of twigs, grasses or weeds and there, standing on tiptoe, turn to face the wind; a spider will then start to let a thread go which the breeze will carry away. When the friction of the air on the line is equal to the weight of the spider and the line, the animal lets go of its support and floats away into the sky. By this means spiders are dispersed over both long and short distances and may even be carried out to sea, where they perish. The white threads floating away in the blue are a common summer sight—'In the year that L. Paulus and Claudius Marcellus were consuls,' says Pliny, 'it rained wool,' which was 216 BC and the wool was undoubtedly a massive flight of spiders, most likely of the linyphiid family, which frequently inhabit walled gardens.

These animals are extremely ingenious in the use of their threads: they will use them as safety lines, carrying a strand of silk about with them, so that they can recover their position after a fall; they will cast them over a large insect struggling in a web to prevent its escape. On sheet webs they will arrange a series of trip lines in order to slow down an insect that is trying to get away and they will also hold a web back with a line and suddenly let it go so as to entangle an insect more securely.

With eight legs, two palps and two or three silk lines to manipulate it is surprising that a spider can be so agile and not become caught in its own snares; they have a remarkable co-ordination of leg movement and can run across a web at great speed, avoiding sticky threads and trip lines in order to reach an insect and give the quietening, paralysing first bite.

The orb webs, those magnificent works of art glistening in the morning dew, are composed of two types of silk and each species of the orb-web builders makes a certain distinctive pattern. The spider squeezes a thread out and lets it float off in the wind; when it catches on an object she pulls the line in tight and then, spinning out a safety line as she goes, in case the attachment of her first line is not too secure, she crosses the first thread, tightens and strengthens it and then proceeds to make the boundaries of the web. This is all done in hard, relatively inelastic thread, and similar filaments are used to make the radiating spokes of the centre and a widely spaced spiral leading to the circumference of the structure. When all is ready, strong and taut, she puts on the evenly spaced series of spiral threads made of sticky silk which is the web proper, and cuts away the temporary spiral threads she first laid down. These last filaments are covered with a coating of gum which draws together to form sticky beads along the length, and these catch and entangle insects. The spider is indefatigable and undertakes enormous tasks. When part of the web is destroyed by a violent encounter the spinner patiently repairs it, and also builds a new web, except for the framework, every day.

In addition to the orb-weavers, there are funnel-weavers, cobweb-weavers or scaffold-builders, triangle-weavers, wolf-spiders and jumping-spiders. The funnel-weavers abound on grass, where the dew can be seen glistening from their webs in the early morning. They construct a sheet supported by the grass, with a funnel-shaped nest at one side in which the spider can hide, or retreat if danger threatens from too large an insect visitor; the end of the funnel is left open so that the spider can escape right away if it is necessary to do so. The owner walks on the surface of this web instead of hanging from the bottom as most spiders do. The cobweb-weavers, or scaffold-web spiders, construct a loose tangle of threads, which may have a sheet half way up the structure. There are usually some vertical threads with sticky drops at the lower end which pull the web down against its tension: when an insect wanders into one of these it becomes stuck and the thread breaks off at ground level, the spring of the web jerks the visitor up and probably entangles it in another sticky thread. The spider can then haul the insect up into the main part of the web, bind it with more lines and dispatch it.

The triangle-weavers make a triangle of threads and sticky webbing attached to a twig at two corners and to a single thread at the third, which is then fastened loosely to a third stem. The spider holds the loose thread, pulls it taut and when an insect blunders into the sticky webbing

Domestic spiders

she lets the thread go so that the triangle contracts and further entangles the victim: she may do this several times before going into the web to feed, and when all is over another trap of the same nature will be constructed.

The jumping-spiders have keen sight and attack by jumping on their victims. They are common on fences and sides of buildings and seem to be examining you very carefully if you stand and look at them. Their jump is amazingly accurate: for instance, they can catch a flying insect by jumping on it, spinning a line as they go, and return to their original position, say on a fence, by means of the life-line spun out. This again shows an astounding co-ordination of eyes, legs, spinners and palps.

Male and female spiders have to be present in considerable numbers on any site if the males are to meet the females, and the species to be continued. The male wanders more or less at random until he finds himself somewhere in the vicinity of a female, when his chemotactic sense will tell him she is near, either from the feel of the ground she has been on, a web she has spun or a thread she has trailed. In some families the male recognizes the female by sight. Dr Bristowe has shown that if in the case of a small linyphiid spider there were only one male and one female per acre on an average the male would have to walk $82\frac{1}{2}$ miles before he found a female: even if there were a hundred of each sex there a male would, on the average, have to walk nearly a mile before he met a female and if both sexes were wandering the distances they would have to travel would be doubled. Eighty-two miles is a quite impossible walk for a spider and even a mile is an enormous distance. In point of fact, spiders are very numerous per acre. Many *Theridon* were present in the field before Bartons End was built; at times there were half a million of this genus per acre and another one and a half million of other kinds of spiders, consequently there was no difficulty in the males being able to meet females. In fact, at that time in England the weight of spiders per acre, taking the average for the country, was considerably greater than the weight of humans per acre. Two million spiders weigh about ninety pounds and four million people distributed over the thirty-seven million acres of England and Wales gave only about $11\frac{1}{2}$ pounds of person per acre, which is an indication of the relative abundance of persons and spiders in Tudor England.

Today some forty-nine million people in England and Wales weigh about 2,363,400 tons, so with the same area now as in Tudor times we get a present figure of 142 pounds of human being per acre, which is well over the weight of spiders on the same acre of meadow.

The food of spiders consists almost exclusively of insects and they will only eat what they themselves have killed. However, hungry spiders will eat insects which would be rejected when more palatable food is abundant. Spiders adapt themselves to the kind of food that is available and their abundance naturally depends on the numbers of insects present in the neighbourhood.

Spiders are able to judge the size of prey caught in a web by the vibrations set up by the struggling insect. They will jerk a radial line both in order to entangle the insect more securely and also to estimate its kind and size. If it is too large or thought to be dangerous, the spider will approach with great care, leave it alone or even cut it out of the web, but if it is something good the animal runs down, killing it at once if small, or entangling it with more silk if large until it is able to approach safely, bite and inject enough poison to paralyse its catch.

Some insects have managed to make themselves so distasteful to spiders that they are left alone even when caught in a web; spiders frequently reject cinnabar moths, wasps, *Sciara* flies and aphids, sometimes after feeling them with their legs or palps, when the chemotactic sense tells them they are unpleasant, and sometimes after actually taking a bite at the insect, when the spider retires to the edge of the web to reject the distasteful morsel: it has every appearance of reflecting, 'These cinnabars make me sick,' literally so. One bite does the moth no harm because it lies still, does not struggle and thus avoids exciting the spider to more activity: in a short while a web-builder will remove the offending but quiet insect from its web and no particular harm will come to it, whereas an insect that continues to struggle will be entangled in more and more silk by the spider, and eventually be killed, even though the animal is distasteful and not eaten.

Strangely enough not all wolf-spiders, or hunting spiders, have good sight: some find their food by wandering around and feeling for insects with their long chemotactic front legs, which enable them rapidly to estimate the nature, size and position of the food and then attack.

Insects have been profoundly affected in their development by spiders' voracity for them, since any mutation which gives an insect prey an advantage over a spider, such as an unpleasant taste or the ability to excrete a disagreeable fluid in an emergency, is most likely to persist in subsequent generations. It is frequently found that certain insects (and similar animals which have succeeded in making themselves distasteful to spiders) which are readily eaten by birds and small mammals (as for instance hedgehogs) are avoided by spiders: examples are lacewing flies, woodlice, many greenfly and millepedes. It seems that natural selection has acted to protect them against their worst and most constant enemy—the spider: but the spiders must not be too successful—if they destroy all the insects in a neighbourhood then they too face future starvation.

Some insects escape spiders because they are very heavily armoured: certain woodlice—though these are crustaceans, related to lobsters and crabs—can roll into a ball and defy any spider: they are protected by plates of chitin as well as a disagreeable taste. Finally other insects have an oily or waxy skin which does not adhere well to the sticky threads of the spider's web and allows the owners to escape easily if caught; examples may be seen in many of the ichneumon flies, which frequently escape from webs.

Spiders destroy enormous numbers of insects, far more than birds do; at Bartons End they certainly slowed down the increase in numbers of the woodworm in the loft, caught a lot of flies and generally reduced the insect population. Although Mary Barton would order the webs dusted away she did regard spiders with a certain amount of awe and superstitious reverence, even fear. They were thought to be poisonous, indeed the Saxon word for spider means 'poison head'. A large spider can give one a bite, but is very unlikely to do so; if it does bite it will inject poison but this is a small matter and not likely to be as uncomfortable as a wasp sting. Mary lost two of her first three children shortly after their birth and half attributed the survival of the fourth to the provision of a little amulet, containing a spider, which her nurse hung round Thomas's neck before he was a week old. As the child throve and became such a joy to his mother she began to lose her fear of the creatures, though she would never allow the webs to remain in corners of the rooms. Consequently the humans in the house must be considered as one of the enemies of spiders.

The worst enemy many spiders have is another spider: there are mimetids which feed exclusively on spiders and these were sometimes found under the thatch. A member of this family will creep gently into the web of an orb-weaver, so as not to be noticed: it will then cut away a few threads to leave space for its special method of hunting. The predatory spider hangs down, draws in its legs and tugs on the lines, apparently giving a signal which the owner associates with the presence of a desirable meal in the web. Down comes the unsuspecting creature: the predator shoots out its long, spiny front legs and traps its host, burying its jaws in a leg or in the body and giving a paralysing or fatal bite, for these predatory spiders possess a powerful and quick-acting poison.

At the end of the mating season the female spiders eat the males, as is well known, though it does not happen in every species: this is a further example of the economy of the spider system. The male spider represents a store of food, particularly of hard-won protein, which would be lost to the race when the males die, as they would in any case at the end of the season. The father literally sacrifices his substance to help secure the future of his offspring. In general males do not live long or eat very much after they have reached maturity. Polygamy is practised and a male, in general, will mate again if he succeeds in escaping from his first adventure. The pattern varies greatly according to the kind of spider: there are some species where a male and female live peaceably together in the same web, though there is no particular constancy among them. The destruction of the male by the female is an example of how, biologically speaking, the female is the more important sex—a fact recognized by humans too. When the ship is sinking does not the captain call 'Women and children first'? Is this chivalry, or the atavistic recognition of the fact that if the race is to survive then the women are the more important part of it?

Among the vertebrates the most destructive animals to spiders are toads, who will feed on large quantities of them: frogs will also eat spiders but not to the same extent as toads. Birds do not consume as many as is sometimes thought, though starlings and tits perhaps eat about five hundred spiders a year each, whereas a toad will eat a thousand and a frog about three hundred.

Mammals destroy spiders by crushing them in the meadows as they walk about, and cattle, sheep and horses consume them as they graze.

Shrews and bats are fond of spiders and man up to about the time of George IV used them in medicines. He, or rather she, also clears away their webs in the house, which causes insects to thrive and spiders to decline.

Insects form the main food of spiders but are revenged on them as a class by parasitizing their attackers. The pimpline insects of the ichneumon family lay eggs in both adults and in their egg sacs. The pimpline settles on a spider and paralyses it with a sting—a thing the spider itself is often doing to insects by means of a bite. It then lays an egg on the forepart of the abdomen where the spider will not be able to remove it when she regains consciousness, in perhaps a quarter of an hour. The egg soon hatches and the young larva sucks the juices of its host: it avoids feeding on any vital part and in fact the spider seems to live a normal life. It spins its web, catches its food, mates and lays eggs but all the time the pimplinid larva is there, always situated so that the spider cannot remove it. A moment comes when the larva is sufficiently grown and is ready for pupation: first it kills its host, presumably by biting a vital spot; it must do this before moving away from the spider or it would in its turn be killed, even though the spider by this time is feeling the effects of the attack, being slow, aimless and making a bad web. The spider dead, the larva leaves the body, finishes eating its erstwhile steed and crawls away along a thread to pupate. It spins a little cigar-shaped brown cocoon; these may sometimes be seen on spiders' webs: from this it emerges as an ichneumon fly after about a week and sets off to look for a spider and start the cycle again.

Other kinds of insects of the same family attack the egg cases of spiders, as do also earwigs, some beetles, anthocorid bugs and a number of other species which are not usually numerous.

The numbers of spiders at Bartons End varied during its history as did the numbers of all other animals in the place, which is illustrated graphically in Figure 11 page 103. The house was constructed in a meadow, disturbing in the process the spiders on one-thirtieth of an acre of grass. 70,000 spiders on this area were destroyed by the removal of the turf and an equal number or more by the tramping of the builders and their animals on the surrounding land as they worked away. The ancients would sometimes sacrifice an animal, even a child, beneath the foundations of a house in order to ensure good fortune, little realizing

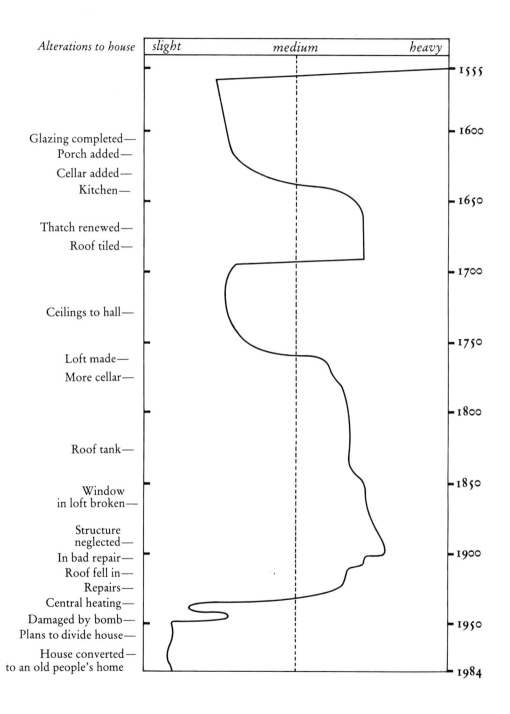

FIGURE II
Population of spiders

that if the sacrifice of an animal were necessary they already had a holocaust of spiders on their hands—in the case of Bartons End it was some 200,000 individuals! The immediate effect of making the building was an enormous reduction in the number of field spiders, such as the sheet-web builders and the hunting spiders; this was followed by a gradual build-up of the population of the orb-web makers and the house-living cobweb-weavers. A porch was added to the house in 1626, with honeysuckle trained round it, which gave excellent shelter for spiders, leading to an increase in their numbers, as did also the digging of the cellar shortly after Elizabeth Barton married Richard Onway (1626). House spiders thrive in damp situations; the drier the house is the fewer spiders there will be in it for two reasons: firstly, the spider is a small animal with a big surface area to weight ratio so that in dry situations it will lose water more rapidly than it can obtain it: secondly, insects are less abundant in dry places, consequently the animals which depend on them for food must also be less numerous.

Up to about 1625 the spider population had been reasonably stable, but with the addition of the porch and then the cellar it began to increase very rapidly. The cellar was an admirable breeding place and added to the dampness of the house, which was increasing as it approached its first centenary firstly because the thatch was now very thick, with the lowest layer somewhat rotten, and secondly, with no gutters to carry away rain-water, a hundred years of splashing on to the wall footings was keeping them almost constantly wet. The cellar was indeed unpleasant, water was often standing there in wet weather and the maids frequently complained of the damp getting into their very bones, only to be told by Mistress Elizabeth that hard work would keep them warm. A trap-door gave access to the cellar from outside which allowed barrels of beer to be moved in and out, but also permitted toads to gain access through a loose shutter, and toads are the inveterate enemies of spiders.

Elizabeth Onway turned her still-room into a kitchen in 1645 and had a fireplace with its chimney-stack constructed on the south end wall. This fire burnt winter and summer: it considerably dried the southern half of the house, with the result that the rapid increase of spiders was checked: they remained at a level we can call 'medium to heavy' for about forty years, in fact until Elizabeth died and her grandson James sold the property to Joseph Munroyd in 1689.

The new owner removed the thatch, replaced it with tiles and fitted guttering to carry off the rain-water, and as a consequence considerably reduced the number of spiders. Large numbers were taken away with the thatch and birds flew in to help themselves to many more. The population remained low, though slightly increasing, for about fifty years, because the guttering stopped the water dripping from the eaves, so that the whole house became much drier. Over the next two hundred years the spiders gradually increased, at some times more rapidly than at others. The first spurt occurred in 1752 when ceilings were made to the bedrooms with a storage loft above. Insects began to thrive in this isolated community, particularly the wood-borers and with them the spiders. Major Hackshaw installed a water tank in the roof in 1822, which increased the humidity there and made life easier for insects and spiders: the pipe runs also gave these animals easier access to the house itself and the accumulation of old boxes, furniture and rubbish made an excellent habitat.

The spiders continued to increase during the time of the two tenants, 1852 to 1899, who were both too busy getting a living from the farm to devote that attention to spotlessness in the house the previous occupants had shown. When the house fell empty in 1899 the fires were extinct, the house grew damper and whilst some insects, such as the cockroach, disappeared, others started to increase, and the spiders would have increased too, had not a small patch of the roof at the north end fallen in during 1903 and allowed the bats to enter the loft space.

The new life in the house fed on the insects and spiders, but they had access only to the loft; in the rest of the house spiders were colonizing the living rooms where they would never have been allowed had any humans been present. Consequently the spider numbers reached a new equilibrium for a brief period at a level just above that of the period 1650 to 1689.

Robert Dunchester bought the house in 1909 and had a grand clean-up, which reduced the spider level another step: in 1925 he installed central heating and this, by further drying the house, rapidly reduced the spider population to just below the level of that of the day the house was first occupied. The Second World War, with its shortages of fuel, caused the numbers to start rising again: the insecticidal treatment against the woodworm almost extinguished the spiders as well, but

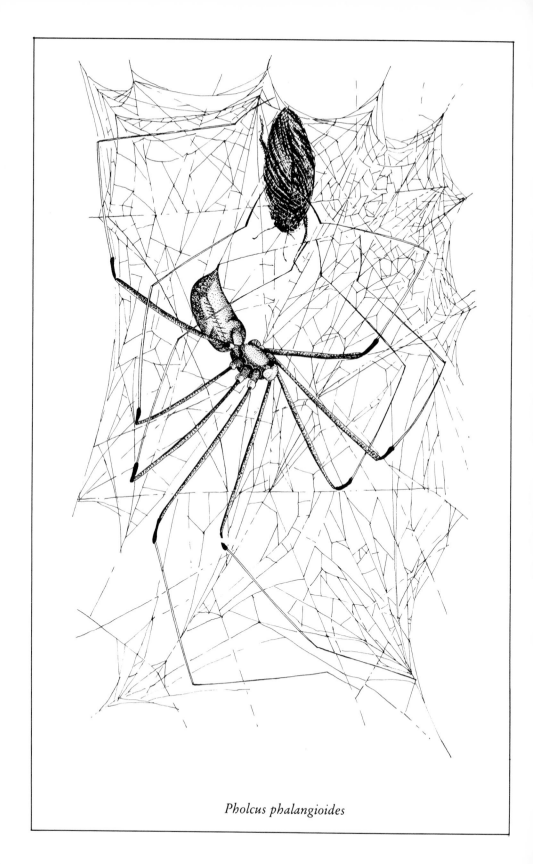

Pholcus phalangioides

recolonization from outside has enabled this Arthropod to retain a footing in the establishment and even to increase again slightly.

The commonest spiders now found at Bartons End are the very long-legged *Pholcus phalangioides* and the long-legged *Tegenaria domestica*.

Pholcus is a very characteristic creature and is not likely to be mistaken for any other species. It lives in houses and out-buildings and tends to be found hanging upside-down in its web, usually in corners at the tops of walls and on the ceiling. *Pholcus* has eight eyes, a body about 9 mm. long and extremely long legs. The carapace (the combined head and thorax) is narrow in front, pale yellow in colour with a central brown marking and some fine hairs; the abdomen is cylindrical in shape and about three times as long as wide, grey in colour and covered with numerous short hairs. The legs are very long, brown and covered with long fine hair, the final joint of which (the tarsus) is subdivided into a number of false joints, which will only be seen by examining the creature with a hand lens. The females may often be found clutching a globular white egg sac about 5 mm. in diameter. It has no particular breeding seasons, for adults may be found at all times.

The *Tegenaria* are also long-legged spiders but do not assume the same characteristic position as the previous species. The body of *Tegenaria domestica* is made up of two ovals and is about 9 mm. long. The forward oval, the carapace, is light brown with faint radiating streaks; sometimes a pair of light lateral bands with a dark border can be seen together with a light central stripe. The abdomen is light brown with only a faint pattern of dark sooty patches. The first pair of legs are about twice as long as the body. A sheet web is built with a tubular refuge at the side and the animal parades on the upper surface of the structure: the web is in fact the familiar cobweb of houses and sheds. This is the creature frequently found in the empty bath, having descended there to seek water and being unable to return up the shiny sides.

The two spiders described above are those that best support the dry warm condition of the modern Bartons End: there are still others present, such as the *Oonops domesticus* which hunts its prey at night by making sudden short rushes at it, whilst a related species, the tiny pink *Oonops pulcher* steals around the webs of other spiders and consumes insects which are too small, or too distasteful to be considered by the web builder, but their numbers are low and are steadily being still further reduced. The large inhabitants breed but little and the population tends to be maintained by colonization from outside: floating on their gossamer balloons these creatures land on the roof and creep upards under the tiles and into the roof space. Now warm and dry, the loft is no longer the favourable environment it was at one time, so that the aeronauts are not particularly successful in establishing new colonies.

During the two periods when the spider colony was at its height—1650 to 1690 and 1850 to 1909—there were a large number of other species represented. Among them may be mentioned the fierce *Ciniflo*, or *Ciniflo ferox*. This is a large creature with a body about 13 mm. long, having a dark carapace and a dusky-grey almost black abdomen: it is a dark sinister animal with pearly-white staring eyes and is responsible for many people being afraid of spiders. *Ciniflo* are cribellates and surround the main threads of their webs with sheet silk combed out by the calamistrum on the hind legs, which gives a bluish tinge to the fresh web.

The mouse-grey spider was common during these times as well. It is a fairly large creature mainly active at night, which builds a silken retreat for itself and its eggs but does not construct a web. The carapace is reddish-brown and the abdomen mouse-grey thickly covered with fine hair so that it resembles a mouse in this respect. The adults are found mostly during the summer months.

During this period the enemies of spiders also abounded and among these were the mimetids, which were mentioned before as feeding on their own kind: they were seldom seen in the house but they attacked spiders in the fields around it and in the garden and thus kept down the number of migrants. In the damp cellar at Bartons End toads kept down the spider numbers, until the boiler for the central heating was installed, the floor concreted and the life there considerably altered. The toads which were found were removed and there was no inducement for them to return.

Chapter 8

Birds

IN THE middle of the sixteenth century, when Bartons End was being built, the bird life of the area was changing considerably, for two reasons. The first was that the great Wealden forests were being cut for the iron smelters and for building houses, and the second was that the waste was being ploughed up as the population grew and more farms were carved out of it. The force and drive of the Elizabethan age—the exploration of America, the increasing population and rising standard of living—demanded more and more of what had been almost a precious metal—iron—for tools, ploughs, cannon, bolts, nails, weapons and so forth, and to supply it and reduce its price more and more of the forests had to be converted into charcoal. The timber houses at first used immense quantities of wood but, as the century advanced, timber began to be used more economically, especially in areas away from the forests; the close-spaced uprights of the early buildings gave place to much wider supports and large rectangles were left to be filled in with laths or brickwork. Timber was even brought in from Norway, so great was the demand. The enormous and wasteful fireplaces also used large quantities of wood and the kitchen fire was never extinguished.

The farm at Bartons End was won from the forest and the waste, the house being built on a meadow which had recently been forest, so that there were two sets of changes introduced here. The woodland birds

gave place to those from forest-edge and meadow and they in their turn
were replaced by the house-using kinds and the birds of the farm.

The Wealden forest abounded in jays and woodpeckers, and as the
woodman's axe cut into the trees, so did the jays and two varieties of
woodpecker retreat, and their numbers diminish, because they could no
longer thrive in the new environment. Though the greater and the lesser
spotted woodpeckers grew fewer, the green woodpecker was able to
stay on and thrive, for the foresters always left a few trees and these were
enough for it. Moreover it enjoyed feeding on the ants in the meadows.

The house at Bartons End was built in the meadow beyond the edge of
the wood. Before this the site abounded with a wide variety of birds.
There were larks, particularly in the shorter grass on the higher ground,
warblers, blackcaps and the coppice birds such as nightingales, blackbirds, thrushes, and chaffinches and tits. Associated with the edges of
the forest were magpies and stock doves, whilst lording it over them all
were the predators. The buzzards abounded then, though now they are
rare birds. Kites were numerous and there were noble peregrine falcons.
The edges of the forest were watched by the sparrow hawks, and kestrels
flourished in the more open country. The hen harrier lived on the moors
and above the forests, and the marsh harrier in the wetter type of
country. The wood pigeon was nothing like so numerous as today;
crows were common and the scavenging raven was much more
abundant, while the parasitic cuckoo was as well known then as now.

John Barton kept hawks; they were the main means of supplying his
table with small game, particularly in winter. He could not hunt the deer
of the forests for they were reserved for the Queen, though there were
many who forgot the royal prerogative and carefully stalked and shot
them with a crossbow—an instrument far more the poacher's friend
than the shotgun, as it was so silent. It is in fact surprising that this cheap
and unobtrusive weapon is not still used!

John Barton kept a pair of goshawks, for as a yeoman farmer he could
not aspire to the aristocratic peregrine. Mary would often accompany
him, and though she might have been permitted a merlin, the 'lady's
hawk', she preferred to have a sparrow hawk, known in those days as the
hawk for the holy-water clerk. They probably caught more game with
these animals than if they had had true falcons, for the action of the
falcons was more noble, more sporting than that of the hawks, but it was

also less generally successful. The falcon, peregrine, hobby, merlin and kestrel, are high fliers, getting above their prey by strong upward flight and then 'stooping' on it, diving at immense speed and striking the bird dead with one blow of the talons, or else grappling with the victim and bearing it to earth. The goshawks and sparrow hawks do not mount much in the air and are less impressive to watch, but they can fly straight at the game at immense speed. It is not necessary to hood the hawks when they are carried about as they only make comparatively short flights, but the falcons must be prevented from seeing birds in the sky as they would always be straining to be away and to rise soaring above them. John Barton had been known to take as many as twenty head of game in a day with his goshawk.

Birds were also caught with nets and many a covey of partridges was taken in this way. All sorts of birds were trapped for the table. Lark pie was a great favourite, these birds being caught by means of limed twigs.

Few birds were shot, because powder was scarce and expensive; moreover, guns were not very accurate for this purpose. When birds were shot they were stalked and fired at sitting, both because of the need to make the best of the scarce gunpowder and because of the difficulty of shooting a bird with a flint-lock gun. There is an appreciable delay between the pulling of the trigger and the discharge of the weapon because the flint must strike a spark from the steel which, in turn, must light the powder in the pan; then the fire in the pan must travel down the touch hole and explode the charge. The noise of the flint striking, and the flash in the pan, would often frighten the game away just soon enough for it to avoid the most carefully aimed gun, and so what with the cost and difficulty of getting powder, and the relatively poor chance of success, it is small wonder that the Bartons relied on nets, hawks and the crossbow for their game.

As both the farm and the farmhouse came into being the bird population of the area gradually changed, and the number of birds that particular place supported was greatly increased. Farming means that more vegetation is being produced per acre than under natural conditions, and particularly that more protein is being made; this in its turn means more animal life can live on that area, whether it be the domesticated cattle and sheep, or the wild animals. The forceful production of vegetation leads insects to increase, and these form the main food of many farm birds. Birds, being so adaptable, soon take advantage of a change in the countryside and in most cases rapidly colonize the farms, but it is because they are such mobile creatures that something must first be said about those on the farm before we come to discussing the ones actually in and on the house.

Crows, except the jays, are typical farm birds, the rook being the most noticeable one, and as Bartons End became settled, the house completed and the routine of the farm established, a rookery soon arose in the tall trees which had been left to shelter the house from the north-east wind; from these high, inaccessible trees the birds haunted the fields and seldom penetrated into the wild country. Jackdaws also liked the farmlands and, though the magpies preferred thick plantations and wooded slopes, they also fed from the farms whenever they could. Starlings soon fed in the open fields and were vey numerous; they liked bare soil or short grass, travelling long distances from their roosts to find it, whilst they avoided the tall crops. Linnets were found on the stubble and goldfinches fed on weeds and thistles, so that where they abounded was usually a rather neglected place near a wild patch. Chaffinches were birds of the farm, garden and orchard. The house sparrow was a creature which rapidly adapted itself to the new conditions and accepted these new animals—the Bartons—and their way of life very eagerly. These birds never travelled far from the house or farmyard, breeding as they did in the buildings. They foraged in the yard, stubbles and fallows and used particularly to follow the horses, feeding on both the droppings and the spilt fodder. The tree sparrows were not nearly so numerous, but were also farm birds. Skylarks abounded on the short pastures. The pied wagtail was a well-known bird there, nesting in the farm buildings and consuming a large volume of insects. Blackbirds and thrushes were common, as were also the dunnock, robin and wren: fieldfares and redwings appeared in the meadows in the winter. Green plovers were found on the plough.

Swallows and house-martins were very intimately associated with the farm, for they nested in the buildings and followed the livestock for the sake of the insects these animals attracted. The barn owl and the tawny owl found much of their food on the farm, destroying a lot of mice and voles in the course of doing so, and the barn owl relied on the farm buildings for its nesting sites.

Owls

Of course many more birds than those mentioned above were and are found at Bartons End, but this list contains those that were particularly encouraged by the making of the place, who have adjusted themselves to farm life and who were found there during its long history.

Birds as a race have always been able to adapt themselves quickly to new circumstances because, though some kinds disappear or become extinct, others take their place. The very existence of birds on this planet is due to their successful adaptation through a long period of evolution. After the amphibians crept out of the sea and conquered the land, during the Devonian age (about three hundred million years ago), the reptiles arose and as they scrambled among the trees over the centuries, the ability to extend a leap by means of a gliding motion gave such animals an advantage over their competitors. Two such forms of gliding mechanism developed—one in which all four limbs formed the basis of wings with skin stretched between them (in rather the same way as in a bat's wing today), and the other where the forelimbs only were used. In this last case, the reptilian scales of the limbs began to change and become feathers; even though this was during the warm Jurassic period (ninety million years ago), it is thought that the original advantage of the change from scales to primitive feathers was the greater warmth of these last, which would be particularly advantageous in the polar regions.

When the colder Cretaceous period followed the Jurassic, the advantage of feathers was yet more pronounced, though the flying reptiles still existed and in a wide range of sizes from a creature no bigger than a sparrow to the enormous pteranodon with a twenty-five-foot wing span. This animal, twice the size of an albatross, lived a life similar to that of the modern bird; their fossils are found in deep chalk which must have been deposited far from land, so we know they ranged far out to sea, fed on fish, rested on the water and only returned to land to breed. This enormous creature slowly flapping its wings, gliding and soaring over the sea must have been a strange sight in that distant age. The wings stretched between the immense forearms to the legs and were quite bare of covering; the relatively tiny body was just an appendage to the wings. These and similar reptiles could not compete with the other line—the primitive birds in command of the new realm—the air—even though they evolved many of the advantageous structures of birds. They had hard beaks and no teeth, a large section of the brain devoted to securing balance and control in the air, and strong, hollow bones containing air instead of marrow, which made them much lighter, but they did *not* have feathers, that very efficient covering which retains warmth.

As the Cretaceous age became colder still and gave way to the Eocene, the flying reptiles disappeared and the true birds triumphed. They had developed feathers which not only kept them warm but gave them great powers of control in the air. A bird is usually a small animal, which again means that it has a relatively large surface area compared to its weight. A young unfledged bird must eat its own weight of food daily in order to maintain life and grow, most of this food being used to maintain its temperature, so that the advantage to a bird of the warmth-retaining feather covering is immense. The feathers trap little pockets of air which, if it cannot move, is a most effective insulator; the reason a bird fluffs up its feathers and looks bigger in cold weather is that it is trying to make as many of these pockets of dead air around it as possible.

A typical feather consists of three parts: the quill, the rachis, the vexillum, containing the barbs and the barbules. The quill or calamus is the part inserted into the skin; it is hollow and the lower part is filled with spongy matter. The hollow part of the quills, mainly from geese, after being hardened in an oven, formed the quill pens of the Bartons and their successors for a long time at Bartons End. The rachis forms the shaft of the feather and is simply the continuation of the quill; it is grooved on one side, which adds to its strength and is filled with a soft pithy substance. To each side of the rachis spreads the web which makes up the vexillum, or vane of the feather. This consists of barbs which are joined together by small hook-like barbules at the top of the feather but not at its base. The structure is a masterpiece of evolutionary development as feathers are both immensely strong and very light. There are naturally a number of different kinds according to the function they have to perform for the animal.

Quill feathers are found in the tail and wings; those attached to the hand bones are the primaries, and those arising from the forearm, which are smaller, are the secondaries. A small thumb in the bones of the wing also carries feathers, which structure is known as the 'bastard wing' and in effect acts as an aerofoil by playing a part in the control of flight. The tail usually carries from ten to twelve strong quill feathers known as vectrices, but up to twenty-four may be found in some birds; the bases of these feathers are covered by more short ones known as 'tail-coverts'. The body is covered by the down feathers and over these are the body feathers or plumulae where the barbs are quite free, having no hooks, or barbules.

The skeleton of a bird is very light and compact: like the early flying reptiles, the bones are 'pneumatic', having hollow air-filled spaces in them. The sternum, or breast-bone, in flying birds is very large, serving for the attachment of the big muscles which move the wings and which form the white breast meat in our domestic fowl. The bird's bill now no longer bears teeth; it contains a hard, horny tongue and the food is conducted first to the crop, then to the gizzard where it is ground up by the action of the strong muscles and the presence of grit and small stones which the bird continually picks up for this purpose. From here the ground-up food passes to the true stomach, gut and anal vent.

The respiratory organs include not only the lungs but also the air spaces in the bones, and the extensive air system in the bird's body allows a quick intake of oxygen into the blood when the animal is in active flight and using up energy very rapidly. A bird has a four-ventricle heart and a blood circulation similar to that of a mammal.

Birds only have one functional ovary, which is surprising as egg-laying can be very active. The yolk is first formed then, moving down the egg-tube, it is covered with the albuminous white and towards the end of the passage receives its hard shell, consisting mostly of carbonate of lime: from here the egg passes to the cloaca, and then out of the vent. The young have a hard knob on the upper half of the beak which enables them to break their way out of the egg. Some birds, such as the domestic fowl, have precocious young which can feed as soon as they hatch, and others have nestlings that have to be fed for some time by their parents.

Birds live in nearly all climates and in any one spot may be residents, migrants or gipsy-migrants (those which move from one district to another in search of food, but have no fixed migration pattern).

Bird life has many strange sides to it; at times they seem to have mysterious powers of doing things for which man would require complicated apparatus, and on other occasions they seem unable to solve a simple problem. The difficulty in understanding the behaviour of birds is that we are human and we try to do so if not in anthropomorphic terms, at least in anthropocentric ones. Man is the most highly developed and thoughtful of animals and we must be on our guard against seeing human behaviour in animals, for in fact, we should look at it the other way round. What we take to be human behaviour in animals may only be animal behaviour in ourselves.

Not the least of these strange powers is that of migration over vast distances and the return to the same nest again the following year. Why does this small house martin go in the first place, and how does it find its way out and back in the second?

Many of the seemingly inexplicable actions of birds may really be due to their quicker reactions to a stimulus and to the fact that, without postulating any mysterious extra sixth sense, birds do have some of the existing senses more acutely developed than man—their hearing is over a bigger range, particularly at the lower frequencies, for they can hear distant gunfire inaudible to us, and their sight is more acute, particularly for distant movement, as is that of many other animals.

Though 'lower animals', birds have two characteristics of higher ones, which are a freedom to move to new conditions, and usually an ability to live under those conditions, for when one considers the range of climates and surfaces, from the tropics to the pole, from the dry desert to the vast ocean, there are very few zones which do not have some bird life. It is the struggle for existence which has forced birds into all possible climates and shaped their seemingly strange ways. These ways all have a survival value and enable the creature either to get its food or rear its young under advantageous conditions. Even so, it is in birds that we see the first appearance of play, not that this may not in reality have a survival value: birds definitely play, chasing each other about in the air, diving and soaring and playing a game with a favourite perch, such as a flagpole or roof, similar to Tom Tiddler's Ground. They seem to be so full of vitality that at times their surplus energy boils over into these games, and while these and other actions appear to have no obvious biological significance, they may in fact be a means of keeping the necessary muscles and nerves up to the mark.

A bird's reactions to a stimulus may be seen to be a mixture of the automatic and the emotional; there is likely in the final analysis to be far more of a straight response to stimuli than a thinking out of a best course of action, because birds do not think in the way that we do. They feel and have emotions, but no ratiocination. For instance, a flock of starlings flying north may all suddenly turn and fly west. Why do they all turn at once? Is there some mysterious power of thought transference, or some signal given by the captain? The explanation may well be a combination of two things: a far keener sense of time intervals than ours, and a simple follow-my-leader reaction. A man's natural speed over the ground is some three to ten miles per hour (though a champion sprinter can do a hundred yards at twenty miles per hour), and his sense of time intervals corresponds with this pace (perhaps this is one of the reasons why he is so dangerous in a motor-car). A bird may well travel at thirty or fifty miles per hour in the air and with a thirty miles per hour following wind it can have a ground speed of eighty miles per hour, consequently its appreciation of time intervals must be so much the more critical if it is to maintain control, avoid collisions, land on its perch accurately, and so forth. A bird may well see in a hundredth of a second something taking a man a tenth of a second or longer. Moreover, it is not burdened with a developed brain and consequently does not have to think out what it will do: it responds almost automatically and at once—perhaps birds, given suitable controls, would make better car drivers than men. Finally, since the bird is smaller than man, the actual distances the signals sent by the brain along the nerves have to travel are that much shorter in the animal; consequently the time interval between sending the signal and resulting action is that much shorter as well. This means that the apparently instantaneous turning of the whole flock of starlings may well be only apparent and that really they are following one after the other but at intervals of time too small for us to appreciate. Such movements are not always unanimous: sometimes a flock will split as it turns, a minority breaking off in one direction, but the gregarious instinct is so strong that this splinter group will soon wheel round and rejoin the main party.

Birds live in the present and while they are keen and lively they have no knowledge or thought of either the past or the future. They will build a nest as a preparation for the future family, but this is an instinctive reaction to certain stimuli, not a thought-out process. It is man's ability to think out a course of action for the future that enables civilization to progress; at the same time, man loses his instincts and consequently the quick automatic responses to outside factors that so often are advantageous to animals.

Migration, the formation of territory, song, flocking or solitary life, commensalism with man, all have a survival value. By migration birds can exploit a much larger territory and a particular food, and consequently support a bigger population of their species than they could without. Apart from considerations of winter climate, the insect-eaters, such as the swallows, swifts and house martins, would not find their food here in winter so that they must either migrate, change their food, or live only in the tropics, where a constant supply of insects could be found.

We may well say 'Why not then live only in the tropics; it must be very pleasant for a bird?' The answer is that the tropics are already full of birds who have become especially adapted to tropical life, apart from the question of whether the tropics are agreeable to birds or not. Bird populations extended their zones by some of them moving from the tropics to temperate and arctic climates, where the competition was less fierce, and the migration pattern fixed into hirundines has enabled them to exploit a special food in the two temperate zones. The struggle for survival has forced on them this pattern of migration over half the world as the best answer to the problem of continuing their existence. They range freely through their territory, which is measured in cubic miles not in acres, snapping up the insects which have fed on the plants and animals of the surface, becoming much attached to man and his farms, because man increases the supply of insects and provides splendid nesting sites with his buildings—far more numerous and much better than the rare cliffs and caves of the natural habitat.

When the food begins to fail then they must go, and the stimulus which drives them to migrate is the shortening day length. They have taken advantage of a particular supply of food, have exploited it to the full and now they must do so again in the similar antipodean conditions across the equator.

Any particular area can only support a certain number of animals and they are mostly all competing one with another for it. Birds will get their living in this area and there will be competition both between species and within the species for the food and nesting sites. In order that the species survive, a mechanism must be brought into play to ensure that the area does not have to support too many birds, because if there are many more than the zone can feed the result will be that all the nestlings

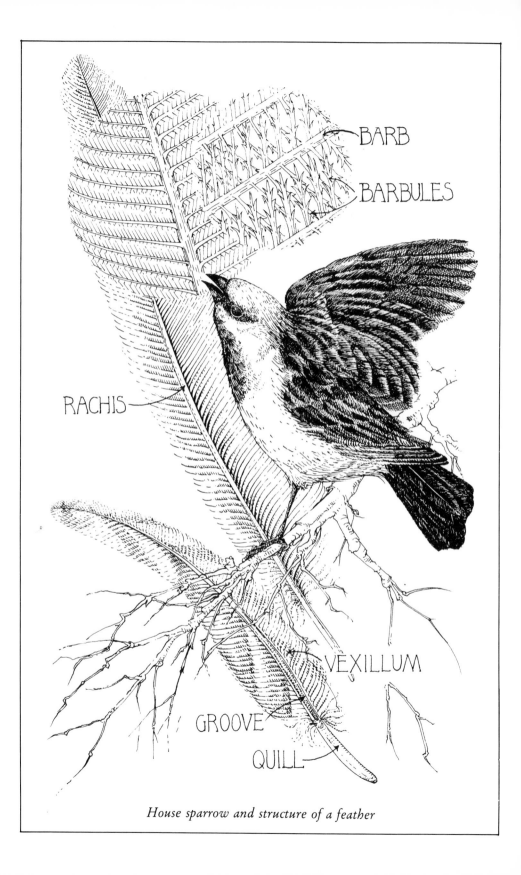

BARB

BARBULES

RACHIS

VEXILLUM

GROOVE

QUILL

House sparrow and structure of a feather

are inadequately fed, do not reach maturity, or grow up as weaklings, and the species will be prejudiced, if not extinguished. The device of territory overcomes this: in the spring the males of many birds establish a zone for themselves, seek a mate and drive out all other birds from their special area.

The purpose of song is to advertise the possession of territory and to warn other birds to keep away. After the male has secured a female, both the birds defend their territory from intrusion by others. Of course the stronger birds secure the best territories and sometimes a strong bird may drive a weak bird out and adopt his territory, or the strong one may carve a territory of his own out of two or three neighbouring areas, but generally once a male has established his territory, he is able to keep it; still more are a mated pair likely to maintain their own private feeding ground as usually they only have to expel solitary intruders. If there is a surplus of birds some will not get territories; they will tend to be the weaker and less enterprising creatures: neither will they get mates nor, of course, breed. This device of territory thus prevents overcrowding, ensures that the area available is used to the best advantage and that the nestlings of all those birds which do manage to secure a territory will have a good chance of being adequately fed.

After the nestlings have flown, the device of territory breaks down, and the birds join up in a social community again.

It is only the smaller passerine birds which form territories in this way and it is an immense change in their lives every spring for they pass from the intense social life of the flock to the solitary one of the territory. The pattern is imposed by the need for survival, for not to conform to it means eventual extinction. In spite of this, however, some birds do not need to make this annual change of habit because they have a bigger, or a more varied food supply, or are strong fliers. The swallows, swifts and similar birds fly so well and range so far in three dimensions that they do not need a private estate but can get their food from, in effect, the public domain—the swifts may easily rise two thousand feet in it.

If it is such an advantage to birds for each pair to have its own private estate we might well ask: Why do they form flocks? To which, of course, the answer is that these flocks also have their advantages. They afford protection against hawks and other enemies as a flock can keep a far better look-out than a solitary bird; moreover, by means of 'mobbing', an enemy, such as an owl or a cat, may be driven off. Flocks also ensure the mixing of different blood lines and are a help in long migration, as often birds will come down and try to help forward a fallen comrade. They also ensure that a newly discovered source of food can be used by a large group rather than by the original discoverer alone, a factor which tends to the benefit of the race as a whole.

The song of birds is often complex: that of the nightingale, which was plentiful at the woodland edge at Bartons End, has about as many tones and sounds in it as has the human voice, but there is no true conversation in birds. The different calls, songs, sounds and notes serve to send signals of alarm, of love, of intention to fly, to return, or do this or that, but there is never any answer given to such a signal, other than a repetition meaning only, in effect, 'message received'. In other words, the song expresses the emotions as well as the definite danger signals; these emotions may be a feeling of well being, of anger, of excitement, of joy at a successful rearing season, or the discovery of a good food supply. They are not languages which can be translated, except by our interjections such as Oh! Hi!, or by the lines of old folk songs like 'Fol-de-loll, fol-de-loll, loll, loll, loll'. Birds will also express boredom by means of song and indeed this is the chief reason why cage-birds sing. Many birds mimic other sounds, including the human voice and these activities seem to be part of their play; they can also recognize the voices of their mates and enemies.

The birds which became intimately associated with the house at Bartons End, as distinct from the farm, were the house sparrows and the house martins—they actually nested in the structure in considerable numbers. Starlings occasionally nested in the creeper on one wall. The chimneys and ridge of the roof were used as perches by starlings, blackbirds and jackdaws.

The house sparrows, who are really grain- and seed-eating animals, were the first birds to appear at Bartons End; never shy of man, in fact dependent on him now for their livelihood, they accompanied the builders on their first day, eating the crumbs that fell from their bread as they took their midday meal, and they have been associated with the farm and the house ever since. At the time of the building the sparrow was found, but it was not particularly common, being relatively scarce because man was scarce at that time too. As John Barton developed his fields and grew big crops of grain and wheat, so did the sparrow population grow too.

As the farm expanded, cats, hawks, limed twigs and netting were all used to attack the sparrow, even to exterminate it, but the creature is amazingly persistent and survived all these threats to its existence. It was called the 'avian rat' in the nineteenth century by Miss Ormerod—William Daker, the second tenant of the farm, a feckless yeoman scholar who preferred to read by the fire rather than work the farm, was a great disciple of this distinguished entomologist. The avian rat was only reduced at the last by a simple ecological change: a reduction in its food supply.

The sparrow population on the farm is now low, at about the same level as when the house was built, because now far less wheat and barley are grown in the Weald; and the horse—that great friend to the sparrow—has now been replaced by the tractor. When the farm was first worked, the sparrows of the house and farmyard numbered about two hundred birds, rising considerably during the great corn-growing epoch of the early nineteenth century when the Hackshaws could have counted some ten times this number, had they set about it.

The sparrows at Bartons End did not have to make territories, as they had a much easier way of life open to them than all the strain of defending their own estates; sparrows only nest where there is a ready supply of food and shelter and Bartons End came well within this category. They got their food easily, and developed a wariness for cats and traps, for it became increasingly difficult to entice them under a sieve propped up with a stick and then pull the stick away as the Bartons had once been able to do very readily. At that time the catching of sparrows was more for the purpose of supplying the table than for the protection of the crops from their depredations. Sparrows were seldom seen singly at the farm, but, except at threshing time, they did not appear in large flocks.

As sparrows get their food relatively easily and have no territory to defend, or migration to undertake, they have considerable time to spend on other activities and it is a commonplace of everyday observation that they fill the time with feeding, bathing, dusting, quarrelling, mating and chirping around with the gang. Over the years they have become more and more dependent on man for, whereas at first they would eat many weed seeds, caterpillars and butterflies, as well as grain, today they depend more on scavenging the rubbish for food, taking the remains in the dog's bowl, or the crumbs off the bird-table, or indulging in similar easy activities. Insects do still form a small part of their diet and, in fact, the young are fed entirely on insects. The fact that the birds no longer eat so many weed seeds is not necessarily bad for the farm, as many of them are not killed by being eaten, so that the birds may disseminate weeds more widely by eating the seeds and then excreting them. Nor is the grain they take necessarily a loss to the farmer, as much of it is taken off the ground where it has shaken out of the ear, or in farmyards where it has been spilt or fallen out of stacks.

Their song was a confused one with a wide variety of call notes and spluttering chirps—not particularly musical—and as the birds were not defending any territories they did not select particularly prominent places from which to proclaim their melodies. They used to start to sing about St Valentine's Day, in accordance with the tradition.

They were strong, ungraceful fliers, merely using flight as a means of locomotion, not for play or for expressing their emotions, and rarely did they rise to any greater height than was necessary to clear an obstacle.

They were fond of pushing their nests into the space between the rafters under the thatched eaves of the house, and were not above using the actual straw of the thatch for the purpose, but mostly they would trail up with long pieces of loose straw found in the farmyard. The nests were well hidden but clumsily made; the greyish-white eggs were from three to five in number with markings in similar tones, or in browns. Usually the first brood was off by May and two more would follow it, although breeding might take place the year round at times.

They showed such variety of marking that it was frequently possible to recognize particular birds and little Prudence Onway, who it will be remembered tamed a bat and named it Master Milton, had one particular sparrow which she called Abigail, that she fed on the window sill. She quarrelled with her brother James when he steadfastly maintained that it was a different sparrow each time.

The first house martin's nest was built under the eaves of the east side in the early summer of 1557, and these birds continued to use this favourable site until the disturbance caused by Joseph Munroyd's removal of the thatch in 1690 drove them away. Five years later they had accepted the change and were back again under the eastern eaves where, at the time of writing, there are six nests.

House martins, swallows and swifts were originally cliff-breeders and while all three have attached themselves to man by using his buildings for nest sites, the house martin, as distinct from the swallow and swift, still uses the old natural sites as well. The colony which established itself at Bartons End was an offshoot of one living on the cliffs at the North Foreland in Thanet. House martins choose their nesting sites in a very arbitrary way—once they have found a place they like, they hold on to it with great tenacity and nests will accumulate there, whilst near by apparently equally good sites will be untouched. Bartons End was one of these favoured places because it was not far from the coast, was near a river, and the farm produced a plenitude of both insects and mud for the nests, though most of this last came from the river. As the house martins were strong fliers and insect-feeders they had no need to maintain territory, so that a colony of nests on a chosen site was more an advantage to them than a handicap.

At Bartons End they spent nearly all the hours of daylight on the wing, not often venturing far from the farm because not only did they like to keep this in view, but also their supply of food was found there. Their territory is a three-dimensional one, in contrast to the sparrows which really had only a two-dimensional zone. The house martins did not exploit such a large volume of air as the swift, perhaps only reaching half the height of these last birds. Swifts range so high and so far that it is sometimes thought that they spend the whole night on the wing, sleeping whilst soaring in the air; even if not true, it is a charming fantasy that this powerful bird should really live in the air as its medium, something in the same way as a fish lives in the sea.

Certainly the male swifts may be seen on a summer evening chasing the hen birds back to their nests and then all getting together, screeching and circling higher and higher till they can no longer be either heard or seen. There are reports of them coming tumbling out of the sky at a mad pace just about sunrise, to dash off in all directions when only a few yards from the ground.

There is a certain amount of communism among the colony of house martins for they enter each other's nests, forage and fly together and enjoy sittting in flocks in warm spots, such as a tiled roof, to enjoy the sunlight. They are peaceful birds, safe from hawks by their speed; and their only real enemies, apart from cats, are the house sparrows which frequently take possession of their nests and even throw out the eggs in them. Many battles with sparrows took place under the eaves of Bartons End, both species of birds screaming and twittering with rage, with varying results, but the house martins were always able to keep most of their own property out of the grasp of these brazen invaders.

The house martins would usually arrive at Bartons End in early May, having come a very long distance—from South Africa—though sometimes they would come in April, or even March. Though they arrived late they would not leave early, as is usually the case with late arrivals, because a late arrival mostly means the bird can only support life in our warmest months. At the farm the house martins would usually start to return to Africa in mid-August, but in some years they stayed on until October or even occasionally till November; they never passed the whole winter in Kent—it would be impossible for them to obtain the necessary food. The same birds would come back year after year until their span of life was over, and start to repair the old nests, or to make new ones if the old had been taken over by sparrows whom they were unable to evict. How did the birds find their way over this enormous distance, and how did this habit arise?

In the first place birds, being free to move where they wish, have a natural tendency to wander; if a hard winter constantly eliminates stay-at-home birds, then those that move away and survive the journey will be the future perpetuators of the species, which will thus tend to have this pattern of migration stamped on it. Then there is the question of distance: from Bartons End house martins, swifts and swallows went to South Africa—a journey of some 6,000 miles, but birds of these species habitually fly enormous distances every day. Swooping up and down, using the air currents in the same way as a glider pilot, the swift does not need to expend much energy in his constant travels from a summer's dawn to the same day's dusk; he can cover forty miles in an hour or easily five hundred miles a day, so that the journey to Drakensberg is put into another proportion when we see five hundred miles as a regular daily beat. We notice the long distances between the two centres and overlook the daily circles and swoops over our heads. The birds will feed as they go if they have the chance; they do not fly particularly high, usually under 2,000 feet, nor very fast, about forty miles per hour being the speed of most of them. They seem to possess a sense of general direction and to follow coastlines whilst avoiding mountain ranges. They may easily have some sense of temperature gradients and thus be guided southwards. They eventually learn to recognize local objects at both ends of their beat. Young birds do not learn the way from old ones, because the young leave first; it is instinct that drives them on, and they must memorize the route so that they can return to the same spot when the time comes. It is noticed that young birds only come back to the same locality, not to the identical nest from which they left.

Instinct—today the biologist is constantly referring to 'instinct': it is a dangerous word and perhaps one that were better abolished, for it is imprecise and thus can be given any shade of meaning desired. When psychologists argue, 'instinct' is frequently thrown in, as being a sufficient and final proof of a favourite theory, according to the arguer's personal convictions. The word 'instinct' arose in the eighteenth century to mark the distinction between the behaviour of man and animal. The age of reason saw man as the ultimate of creation and as activated by God-given reason; by contrast the animal was moved by an equally God-given but very different concept—'instinct'—which enabled animals to do things without first learning how to do them. There seems no occasion today to make this very fundamental distinction, whether one adheres to the theories of Darwin, Lamarck or Lysenko: there are few biologists—or indeed laymen—who would consider man as anything fundamentally different from the rest of creation. The same basic laws of behaviour should apply to him as to animals (though the vice of scientists is to find universal laws), and consequently we can do without the concept of 'instinct' as an animal-held, mysterious, inexplicable thing.

I must point out, however, that quite recently Ronald Fletcher has come to the defence of the word 'instinct' in his book *Instinct in Man*. There is nothing wrong with the word provided we accept a definite meaning for it. Fletcher describes instincts as the 'trains of unlearned behaviour . . . which are activated in a co-ordinated manner when the animal encounters the various situations of its environment . . . instinct is simply a concept used to denote a certain correlation of these features'.

Mankind has instincts: for instance a baby knows how to suck milk as soon as he is born, though he must be taught how to handle solid food. In man most behaviour is not conditioned by these inherited tropisms, but by the higher brain centres—by the process of thought. His instinctive behaviour is over-ridden by a thoughtful one. The same is true in a greater or lesser degree with animals. Some, such as most insects, are almost completely controlled by the inherited pattern, others show increasing degrees of independence from this pattern, until we reach man who has comparatively little instinctive behaviour left in him and is, or should be, all thought-controlled.

The chirruping of the male crickets irresistibly draws the females to them from far away away in the fields: the song of the sirens was equally irresistible to Odysseus and his men, but his higher thought processes enabled him to realize the danger of the song and to overcome this by having himself tied to the mast and filling his crew's ears with wax. The crickets come to the sound even if it means falling into a trap. The humans can take steps to avoid danger, because they can overcome their tropisms with a thought-out course of action.

At Bartons End the house martins would fly down to the River Beult to select the mud for their nests with great care, for it had to mould properly, to dry without cracking and to be strong. The foundation must be particularly well laid and once this was done the rest followed easily, the birds using pieces of straw to give the structure strength. Nest construction was only done in the morning, for each advancement must be left to dry and harden: if too much were built the wet clay would pull itself down by its own weight. When sufficient mud for the day had been put on, the birds were free to feed and amuse themselves, which they did with great expressions of joy and pleasure. The nests were the size, shape and colour of a coconut shell with the top cut off, placed high up on the wall so that the overhang of the thatch protected the entrance. The eggs were white, three to five in number and two broods were usually raised with sometimes a third, both the cock and the hen sitting on the eggs in turns. The older of the young birds helped in the feeding of the subsequent broods and the fledglings were not only fed but also had their excrement, a tough kind of jelly, carried away by their parents and siblings. When three weeks old the fledglings became ready to fly and for a short time they were fed on the wing.

The house martin has its underparts and rump pure white and the

House martin at nest

upper parts of the head to the tips of the wings are blue-black turning to dark brown, which is much browner in the young birds: in contrast to the swallows the feathers come down their legs as far as the toes. They are about five inches long with a much blunter tail than the swallow. Their song is but a mellow twittering.

The Bartons were at a loss to account for what happened to the swallows and martins in the winter, for they had no idea that migration took place at all. There was a theory that they massed and went underground where they might occasionally be dug up. Dr Johnson, who was much admired by the Munroyds, gave currency to these ideas and invented a wonderful word to express them when he wrote: 'swallows certainly sleep all the winter. A number of them conglobulate together by flying round and round, and then all in a heap throw themselves under water and lie in the bed of a river.' The belief persisted for a long while; in fact, until people found the same birds in other parts of the world and noted flocks of them winging their way north and south to follow the sun. Finally, putting marked rings on birds, particularly the fledglings, has led to definite knowledge of their movements.

House martins are much attacked by parasites, as we shall have occasion to notice later on, so that it is not necessarily an unkindness to birds to destroy the old nests because these may well contain the eggs of mites, bugs and lice which take a heavy toll of the bird's blood. The house martins were always welcome at Bartons End: that they made the place lucky was accepted by all and the pleasure that Joseph Munroyd took in his new fashionable roof was not really complete until the house martins came back to it.

It was not realized at that time, or even at all widely now, that the house sparrows were inveterate enemies of the house martins as these cocky little ubiquitous creatures were always seizing or disputing the noble fliers' nests. In the first ten years of this century when the house was empty the sparrow population only fell a very little because horses were still being used on the neighbouring farm and the martins held up their numbers. Today they are somewhat threatened because, though the sparrows decline, there being no horses on the farm and hardly any grain in the district, which gives the martins a chance to thrive, yet there are factors operating against the martins as well. Myfanwy Thompson and her children love all their birds and particularly their lucky house martins, for the tradition has survived. They put out crumbs for birds in the winter which mostly benefits the house sparrows, draws them to the building where they occupy the martins' nests, whilst these are the other side of the world. The new farmyard is now so much cleaner: the dung is stacked scientifically so as not to breed flies, the cowsheds are sprayed to kill any that do appear, and all pests of the farm crops are now kept under control, with the result that there is not the same insect abundance in the neighbourhood as formerly, meaning less food for the swallows and house martins. The new farming keeps down rats which are enemies of birds; they may get access to the young of swallows but rarely can reach the martins' nests high on outside walls, so the decline of the rats was not much to the advantage of these last birds. Bartons End may well lose its martins unless the humans harden their hearts against the winter sparrows.

The house sparrows and the house martins were the only birds that lived in or on the house, but many other birds used it. The tits found the thatch a source of insect food and would pull it out to get at these creatures; their depredations were the reason Joseph Munroyd gave his friends for getting rid of the thatch and replacing it with tiles, though the real reason was to improve his status in the county, as indeed it did, for he soon became a magistrate, and his grandson married a wealthy woman. When assessing the effect of one animal or another, psychological as well as physiological factors must be taken into account. The tits were destroying insects living in the thatch but these creatures were not directly feeding on the straw itself but only on psocids and similar insects living on the decaying fibre; consequently by pulling out the straw the tits were harming man, as this allowed more rain to enter and allowed the wet condition to spread. By providing the motive for a new roof the status of the human inhabitants was improved because a tile roof is safer than a thatch one—it will not so easily catch fire—and does not need renewal every ten or twenty years; moreover, tiles are not so easily torn off by storms as thatch is.

Birds of many kinds were regularly to be seen on the roof ridge. Over the years they gradually adapted themselves to these new structures and the black redstarts now prefer perching and singing on the house to any tree in the neighbourhood. Blackbirds and jackdaws also see in the building a splendid substitute for rocks and crags and they were frequently present on the ridge of the house. Jackdaws would at times make their nests in the chimneys and cause a lot of trouble when the winter fires came to be lit. Every now and then a robin would be found trying to force its way in through a shut window. This was usually in a rather dark room and the bird could see its reflection, which he took for an intruder into his territory and so tried to expel him, an experience which baffled him considerably. Glass is a phenomenon birds have not yet mastered, perhaps because their encounters with it are rare. Knowing how to deal with glass may have some survival value, but so few birds meet this particular problem that no instinctive knowledge of this subject has yet been acquired.

Chapter 9

The Dry-litter Community

UNDER NATURAL conditions there is a rare group of animals belonging to the dry-litter habitat—rare because of the scarcity of dry litter in nature. Wet litter is very common, particularly in the woods, at the foot of cliffs and so forth where rotting vegetation, dung and animal remains may be found, together with its own special fauna—earthworms, woodlice, springtails, mites, millepedes and so forth—living in and on these wet-litter substances. By contrast, dry-litter habitats under natural conditions are found only in the nests of certain birds, such as those placed on cliffs or in dry caves, or in a few dry deposits of dung from bats or birds (all very rare compared to the wet-litter sites) but nevertheless containing their own special fauna.

The dry litter tends to consist of high-protein material such as waste hair, feathers or fur or dung having a high nitrogen content, and the creatures living in it, or some of them, were first of all predators on other small animals and then became scavengers of such discarded materials of higher animals as moulted feathers and hair, cast snake-skins and dung. They lived a highly specialized life within their unusual realm and had to perfect devices which would economize water, the scarcest material in their chosen environment.

When man came on the scene he had a considerable effect on these rare animals, because he was a great creator of dry litter; he liked dry places—caves, huts, castles, houses—which he would keep warm and, to begin with, not particularly clean, so that dry litter began to collect in depth and a vast new dry-litter habitat was created. Even though he was not covered with fur, did not moult or cast his skin and left his dung in the wet-litter habitat, yet he started to wear clothes and scattered bones with meat still adhering to them about the floors of his caves, huts and castles. His first clothes (skins and fur) were to those dry-litter animals a huge supply of their chosen medium and as man began to weave coverings of wool and keep them in chests and cupboards, the animals began to move out of the birds' nests and bat dung and take to clothing as their main means of livelihood.

As man moved forward to improve his world, so he influenced the life of many other creatures; he greatly increased the opportunities and numbers of these erstwhile scavengers, a step which caused considerable annoyance to their hosts. They have not, however, even now abandoned their old sphere and are still found in birds' nests and deposits of dry droppings, and frequently it is from these sources that the creatures spread into man's stores of food and clothing.

What we now call the clothes moths are the principal members of this community. There are in Britain some five different kinds of moth that will feed on this new dry-litter field. Three are clothes moths and two are house moths: the three are the common, the case-bearer and the white-tip clothes moth, and the two house moths are the brown, or false clothes moth, and the white-shouldered house moth. To these must be added the spider beetle, carpet beetle or woolly-bear, which operates in much the same way as the clothes moths. The furniture mite is another animal which belongs to this group, and then there are many more who have specialized in sharing man's stores of food, of which the flour beetle was the commonest when Bartons End was built, but to which others, such as the Indian meal moth, the grain and rice weevils, the fruit moths and the bookworm have been added as trade and commerce interchanged both animals and goods from one part of the world to another.

These were the main dry-litter creatures that fed on man's goods in the house, but they were not the whole dry-litter community for there were a number of other creatures which benefited from these sheltered conditions. Silverfish gradually developed in the house, but did not do much damage; they are the most primitive of all insects and can be described as living fossils, though very minute ones. The Bryobia mite or grass spider frequently came into the house to pass the winter and was fond of the thatch, the lower and upper layers of which formed part of the dry-litter habitat, particularly by 1660 when six layers of new thatching straw had been laid on the roof without the old thatch ever having been completely removed. Here mites, psocids, silverfish, springtails, fleas, Protura and even woodlice abounded, but much of this old thatch was moist because it was in bad condition, being so thick that it never fully dried out, and consequently not all the thatch can be considered as the dry-litter world; moreover, it was not a high-protein dry area, but a comparatively low-protein, high-carbohydrate zone. Many of the scavengers here were actually living on moulds growing on the straw rather than on the straw itself. Strangely enough the top layer of thatch is more often dry than wet because the wind or the sun dries it so quickly after rain, but the variation makes it unattractive to animals as a permanent habitat.

The driest area of thatch was around the vast chimney going through the north end of the house, where the kitchen fire was never extinguished and the parlour fire blazed all the winter. The Bartons and their contemporaries well knew the danger of fire in the thatch—rakes and forks were always kept at hand for the purpose of throwing off flaming thatch from a roof—and the area around the chimney was treated with a mixture of lime and salt in order to reduce the fire risk. It was very effective— at least there never was a fire at Bartons End—but it made this zone very alkaline and discouraged the dry-litter animals from settling there.

Clothes moths were well known to the Elizabethans; their clothing was of wool, linen and leather with a little silk provided for the most wealthy, and all these except the linen would be attacked by the clothes moths. Brushing, airing in the sun, putting the clothes away clean and laying lavender and laurel in the clothes closets were the methods Mary Barton used to combat the ravages of these insects. They were effective too (except perhaps the lavender, though this may have some slight repellent action on the adult moths) and remain so to this day.

Through the centuries as the house was improved so was it kept that much warmer and so did the clothes moths thrive increasingly, but now that the present owners have the place centrally heated there is another crisis in the lives of the clothes moths because, though they like the heat, it is becoming too dry for them. Moreover moth-proof carpets and clothing are an additional brake on an increase in their numbers, as are also artificial fibres such as nylon and terylene. Neither are modern manufactured woollens all that a clothes moth wants, for it is found that it will breed much more quickly on raw wool which obviously contains some vital substance in relative abundance, compared with the manufactured article. If fabrics are clean they can live on them but they prefer soiled material containing sweat, food or excrement which enables them to obtain supplies of vital elements (possibly certain fats, vitamins and amino-acids) more readily than from clean fabric alone. This is one of the reasons why putting clothes away clean helps to protect them from moth.

Finally, moth powders and sprays based on modern insecticides are a serious threat to their continued survival as a race. Together with their enemies, the spiders, they are declining, though by no means extinct as yet. They have overcome many setbacks in arriving at their present way of life and, breeding as fast as they do, they may yet overcome these new difficulties.

Here again we can see the struggle for survival imposing this pattern of life on the clothes moth. A medium containing food material, protein, carbohydrate, vitamins and water, albeit this last was comparatively scarce, offered itself and was duly colonized. To do so the creatures had to acquire some unusual and special habits, for they had to be able to digest the fibrous proteins found in wool, hair and leather such as keratin, fibroin and so forth, which are digested by few animals. Nevertheless they have done so; they had to, or else leave the field untouched. Measurements have shown that about half of the fabric consumed is actually used by these animals and the rest excreted as waste. It is a large percentage in view of the difficult nature of the medium; food passes through the gut very slowly because it is so hard to digest. If a larva is first starved for a while and then fed, it will be two days before anything is excreted.

The creatures can readily obtain oxygen from the air and have a large surface area compared to their weight, which makes this easy as they breathe through pores in their skin (spiracles) not by means of lung books. The larvae live on or in the medium the whole time so the question of water supply is important; as they breathe they lose water; they obtain very little with their food for wool and hair do not contain much of this substance so essential to life; to overcome this difficulty they have acquired the ability to absorb water directly from damp or dampish air. By this means their vital supply of water is maintained. They are able to vary the length of their larval life considerably, which fact also has a notable survival value. When food of the right kind is scarce or water cannot be obtained either from the food or the air, the larva can rest or develop more slowly and into a much smaller adult, perhaps a tenth of the weight of a normal insect, though still, if a female, able to lay fertile and viable eggs.

The drying of the atmosphere of the whole house by the central heating system is a big threat to the way of life of the clothes moth and death watch beetle, for it makes the question of water supply acute; on the other hand, it does provide warm conditions all the winter, thus shortening the life-cycle and enabling the generations to succeed each other more rapidly, which causes their numbers to build up again. The balance between increase and decrease is a narrow one: very small things may make all the difference to the success or failure of an animal. However, clothes moths are not very likely to become extinct at Bartons End as long as birds are encouraged there; while sparrows can force their nests under the guttering or eaves and martins build on the walls, there will always be a supply of these creatures ready to find their way into the house, with its vast supply of dry litter. As soon as a dry-litter animal finds itself in a favourable position in a house its numbers will build up very quickly, similarly the numbers will decrease equally rapidly as soon as measures are taken against it, but in spite of these the birds' nests, together with antique furniture (frequently with old baize or stuffing in it, full of moth) form a constant source of material for the renewal of the population.

The first of the clothes moths at Bartons End was the common clothes moth found in the first sparrows' nests under the thatch of the eaves. The tiny eggs of this creature are oval and ivory-white in colour; found singly or in groups of two or three, they are laid on the surface of tight-woven fabrics or between the strands of wool in blankets and looser woven cloths, or at the base of hairs in furs. Some fifty eggs are deposited

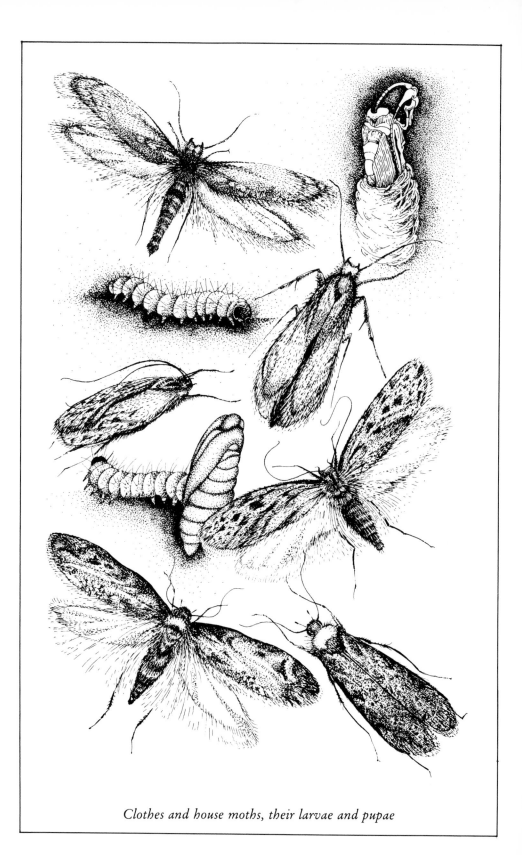

Clothes and house moths, their larvae and pupae

by each female, but the quantity naturally depends on how well fed the moth was before pupating. After laying the eggs the female dies; the eggs hatch in a week or ten days' time. The larvae are white with yellowish-tawny heads and are about ⅔ inch long when fully grown. As soon as they hatch they begin to feed on the 'dry litter' on which the eggs have been laid. When faced with mixtures of animal and vegetable fibres the moths seek out and consume only the animal ones. The material if coloured shows through the skin and makes the larva inconspicuous, a fact having a considerable survival value. Many of the maids and mistresses at Bartons End were not sharp-eyed enough to notice the beginnings of an attack on the blankets, jerkins, hose and multitudinous succession of garments in the long history of the house, and so gave the tiny, vulnerable caterpillars opportunity to develop to maturity.

The common clothes moth larva usually builds a tube from silk and the fibres of the food medium on which it has been feeding at the time, projecting its head from either end to eat. When a new area is to be browsed the larva either extends the tube or abandons it and constructs another one. Not all larvae do this; a few are free-feeding over the area, or just spin a little web silk to protect themselves. Here again is posed a question as to the balance of advantages to the race. The tube protects the larva against certain parasites but renders it more conspicuous to man, while the free-ranging larva is less conspicuous to him but more exposed to attack from ichneumon and braconid parasites. Which is the greater danger?

The noise of the moth feeding, when amplified by microphone, is very startling, a vast tearing and rending of fibres repeated in a rhythm of three or four bites followed by an interval of two or three seconds' silence, and then repeated. It is indeed fortunate for the moth that our unaided ear cannot pick up this sound, for such an alarming noise would not only advertise the presence of the moth but lead to instant defensive action as well! The larvae of the common clothes moth by no means confine their feeding to wool and fur. At Bartons End poisoned mice sometimes died behind the wainscots and these were attacked by a number of insects, including the clothes moth. They proved a valuable source of moisture to them, for not only was the fur used but also the drying but still moist flesh on the mouse bones was colonized; however, this attractive source of food also had its dangers for the moths, for those that attacked the stomach and viscera were killed by the same arsenical or phosphorus poison which killed the mice. The moths attacking the dead mice helped destroy the body more quickly and were advantageous to man in this respect. The action of the humans at Bartons End to reduce one animal obnoxious to them had the effect of increasing the opportunities for development of another equally objectionable.

The moths also occasionally made experiments in living on purely vegetable materials. In the early days of the house, they successfully managed to develop on a few porridge oats lying neglected in a sack, and on some crushed barley fallen into a corner in the brew cellar; however, in these cases they always concentrated on the high-protein portion of the vegetable material. They also attacked silk and leather.

The ancestors of this moth seem to have lived as predators on the larvae of ticks, mites and similar creatures and from this to have adopted the scavenging way of life as being easier; they then pushed a branch out to attack man's goods. It appears a most adaptable animal and may well have genes in its make-up which will again face man with a new development—nylon may perhaps fall into its power—or the ultimate threat (imagined by Stephen Vincent Benét of a termite): one day an entomologist may find a clothes moth larva starting quietly to eat into a steel girder. It has already attacked the insulation of telephone cables in London.

After the larva comes the pupa; first a silken cocoon is spun, and attached to it are threads of the material on which it has been feeding. This economizes on the silk it has to spin, leaving more food reserves for the succeeding adult and the future generation, and is also an aid to concealment. The pupal case is formed within the cocoon, and the duration of this stage is again very variable, lasting from two to six weeks, a lower temperature prolonging the process. At the end of the due time, the pupal case projects from the cocoon, splits at the end and the moth emerges.

The newly-hatched moths are a bright golden-buff colour, the wings free from markings and with rather loose scales, some of which easily become detached. The wing expanse is about half an inch, the males being a little smaller than the females. The female is full of eggs and rarely flies as she is so heavy; moths seen on the wing are nearly always males, or spent females who have already laid their eggs and are thus light enough to fly. The humans always chased and killed any small moth seen fluttering in the house, which did very little to keep down the population of clothes moths, because some were garden moths drawn to the lights at night or, if clothes moths, were the males, or empty females who were about to die in any case. Of course, if enough males could be killed the future population would be affected, but killing a few males will make very little difference to the numbers of the next generation, a fact which is true of most of the animal kingdom and is no doubt the reason why only men used to go to war and women did not. The moths, particularly the females, try to escape danger by running for shelter in quick, dodging, characteristic dashes, though sometimes they take short flights as an additional measure.

The length of life as an adult is again very variable; the males live longer than the females, the former averaging about four weeks and the latter half this time, or a little more; but in cold winter weather females will live some five weeks. The adult moths tend to be abundant in early summer and in the autumn, but now at Bartons End they may be met with all the year round, due to the central heating, though not in the large numbers that were known in the past.

Much was the damage and many were the minor tragedies that these moths caused the humans at the house over the four hundred years of its life. When young Elizabeth Barton (the last of the Bartons) married Richard Onway in November 1626, her grandmother Elizabeth arranged a magnificent celebration for her in the old house; telling her she would eventually inherit the property and that Richard Onway was a lucky young man, she gave her granddaughter the key of her old chest, saying, 'Take it, child, for if today you journey to Canterbury you must needs keep warm, but for that purpose your blushes will not serve so well as my fur mantle.' 'Oh, mistress,' said Martha her maid, 'it has not been open for five years!' Nor had it; when the chest was opened the result can be imagined—the fur was ruined by an attack of moth. The old lady's temper was not improved by the knowledge that she herself had always refused to allow the chest to be opened and though the unfortunate maid tried not to look at her mistress the thought was much in the air. With the help of another girl she quickly thrust the remains into a basket and plunging to the bottom of the chest brought out a silver plate, christening mug and a fine lace veil. 'Mistress,' she said, 'here is the christening cup of your dear father, Master Ephrael Standish, and your point lace veil and a plate that Mistress Elizabeth will ever cherish, and I will cut out the fur and make up a cover for the cradle of their first-born, and all will be well.'

The bride quickly seized these articles and admired them; the situation was saved. It was a lesson she never forgot though she lived to a ripe old age, ruling the family with a rod of iron to the last.

The incident was not unlike a custom of those days when certain things, usually unpleasant, were done to impress the minds of young people at some important event, in order that they should not forget it. A fourteen-year-old boy might be asked to witness a will or a contract and then be beaten or thrown into a pond, for in remembering this undeserved punishment he would also remember the will or the contract. 'Beating the bounds' of the parish had the same motive, but it was not the bounds that were beaten, but the choristers at the main points on the boundary, it being hoped that as they aged they would never forget either. It was a custom that could have caused parochial poaching of land and choristers, had one parish made it known that they did not beat choristers who forgot the boundaries!

Not only did Elizabeth (now Onway) remember the danger of moth (and the Bible mentions it as well), but the happening also taught her that she must always at least consider the opinions of others, though she might not accept them. It was a turn of events which brought prosperity to the farm, for many new ideas were adopted to the great benefit of the family.

Small events have larger consequences; the moths had crept into the chest through an ill-fitting lid and a warped board and prospered in a comparatively isolated world. Their very success nearly spelled their doom through the next ninety moth-generations or more, for Elizabeth Onway never forgot to order the common precautions against moth, which were mentioned above. Elizabeth made one important observation on this scheme which was a little in advance of her time, considering her social position and her sex.

It was generally concluded in those days that insects and creepy-crawlies in general were spontaneously created out of 'corruption', that rotting meat engendered maggots; dung, flies and shut-up clothes, moths. Elizabeth accepted this to some extent, but noted first that the clothes were not rotting and, secondly, that the maggots in the clothes turned to moths; hence she argued the moths could turn to maggots. There was even some confusion over the name, for the word 'moath' had once meant the maggot, but now meant the flying creature, which to her suggested a close association between them. Ever practical, she decided to try keeping the moths away from the clothes by having a chest made with a very tight-fitting lid; fortunately the clothes she put into it were free of eggs and larvae, so her theory was triumphantly vindicated, but one of her daughters was not so successful with 'mother's moth-proof chest', which she left full of clothes unopened for two years, because moth eggs had gone into the chest on the clothes when first put there and throve in their closed community, for it happened that no parasites had gone in with them, nor could they get in, to restore the balance. Mother's moth-proof chest became a family joke.

Elizabeth Onway's ideas were in the air. William Harvey, the king's physician, published his book on the engendering of animals thirty-seven years before Elizabeth died, and his dictum 'Omne vivum ex ovo' had appeared in one of the news-letters which they received from London from time to time and had greatly intrigued her, though they did seem to flout The Old Testament, for did not The Book say 'Out of the lion came forth sweetness'?

How often at Bartons End did the humans go to their stores to take out a special garment and find it moth-eaten! What feverish last-minute repairs were made; what firm resolutions were passed never to let it happen again, only to be broken and the cycle to repeat itself. The moth population grew to its height in the last half of the nineteenth century: this was the period of the two tenancies. Both men had a hard struggle to make the farm pay; the first, Emmanuel Burrows, because of the depression of prices, and the second, William Daker, because he paid very little attention to the farm, leaving the work to be done by his relations, whilst he read books on how things should be done.

The moth population built up then because anything unwanted was pushed into a corner, or into the loft, a minimum of cleaning was done and all the scavenging insects throve. When the last tenant left in 1899 the moths were in no small degree responsible for the dirty state of the house. Even if at first Oscar Hackshaw had not wanted the place for himself, it would have been difficult to sell until it had at least been well cleaned and preferably renovated as well, so that Oscar was fortunate in that a neighbouring farmer desired to add the farm to his own property and offered a good price for it.

The house remained empty of man for ten years, and the moth population fell very rapidly. It was no longer warm and the almost inexhaustible supply of dry litter continually provided by man ceased to come forward, while that left by the Dakers was soon exhausted. The population of the common clothes moth was almost nil in 1909, the year Robert and Myfanwy Dunchester bought the place; only a few were left in that constant source of material—the birds' nests.

Because the clothes moths have proved such a continuous nuisance to man, they have had one far-reaching effect on him, for the insecticide DDT was discovered in 1938, whilst Dr Müller, a Swiss chemist, was looking for new moth-proofing agents for fabrics. DDT is such a powerful insecticide and can kill so many disease-bearing insects (malaria mosquitoes, plague fleas, typhus lice, to name a few) that millions, who would in the ordinary way be dead from these causes, are now alive and healthy.

Would DDT have been discovered so soon had the clothes moth not been so persistent? Taking this into account we can perhaps say that the moth has been beneficial to man, for it has kept down many murderous diseases. In the Second World War, for the first time in the history of war, more people were killed by bombs and bullets than by disease.

Other clothes moths were also found at Bartons End, though not so numerous as the 'common' variety just discussed. The case-bearer clothes moth is very similar to the common moth and has adapted the habit of the feeding tube to the point where this article forms a permanent home, which is carried around by the larva. The case is made of silk and pieces of the food material (in this respect similar to the feeding tube of the common clothes moth). It is broader in the middle than at the ends which allows the larva to turn round and feed from either end at will; rarely is the insect able to live without its case. It is white, four-tenths of an inch long and has a dark head. It feeds in much the same way as the common clothes moth but can adapt itself to feeding on a stranger variety of vegetable products than can the common moth. Dried aconite root, cayenne pepper, horseradish, mustard seed, ginger and similar drugs and spices it seems to relish, as well as the more common high-protein vegetable foods used by its cousin, the common clothes moth.

The larva pupates inside its case and usually the adult does not emerge from this until the following spring. The moths themselves are mainly found from June to October, though there is much overlapping of generations; they are about the same size as the common clothes moth. The wings themselves—the forewings—are dusky brown with three well-marked spots on each. The male moths usually hold their antennae erect, whilst the females fold them flat along the wings. They shun light when disturbed and run to cover rather than fly, in much the same way as the common clothes moth.

The third clothes moth, the white-tip clothes moth, was only occasionally found in the house, though it was fairly common in the stables and harness room of the farm, as it has a definite preference for feathers and skins, especially if these are raw. It is sometimes called the tapestry moth, for when found in houses it is frequently in these hangings. Horse collars and the hair stuffing of carriages are much attacked by this creature.

The adult moths have a wing span of about three-quarters of an inch, the males again being smaller than the females. The head is white, as is also the base of the forewings; the lower halves of the wings are white, sometimes speckled with black. When the moth is at rest it has a protective coloration suggesting a bird dropping, which points to the fact that it is a bird-associated insect and has turned to man's 'dry litter' comparatively recently. The larvae are similar to, though rather bigger than, those of the common and case-bearer clothes moths. Both this creature and the case-bearer only throve under the crowded and strained conditions in which Emmanuel Burrows and William Daker lived. Bartons End does not know them at present except for a stray introduction now and then from a bird's nest, or from an occasional one flying in from the farm buildings.

The brown house moth entered Bartons End as a scavenger of vegetable remains, then turned to man's vegetable food and finally to his animal products. The adult female has a wing span of about an inch, the male again being rather smaller. The forewings are dark brown in colour and have blackish spots on each, two together near the centre and the third half way between this and the wing tips; a number of small black dots are found at the end of the wing. The general outline of the moth when at rest is much broader and more oval than that of the pointed or narrow clothes moths.

The fully grown larva is about three-quarters of an inch in length, of a shining white colour with a tawny-yellow head. It constructs no feeding tube, but moves around freely in its medium, occasionally spinning a little silk webbing.

The brown house moth first established itself in the thatch at Bartons End from whence it passed to the hall where it lived on straw, chaff, rushes and waste grain left in old sacks and skeps. Once in the house, it turned to such food as peas, oats and barley, and any dead insects it came across.

From these successful beginnings it turned towards the drier litter in the house, attacking the stuffing of the chairs, furs and the leather bindings of books. When it attacked books it was those on the lower shelves which were the first to suffer, for it rather liked damp situations. Consequently it was very much at home in the cellar which was always damp until the boiler was installed there. Its wide feeding range is capable of causing much damage to man.

It caused Major Hackshaw very considerable annoyance, in fact at one point he was prepared to abandon the farm. The brown house moth, as we have seen, was in the cellar, and when the half-pay major purchased the farm in 1818 he moved his most valued spoils of war into it—two pipes of port, a gift from the grateful inhabitants of Portugal who had been most anxious to show their appreciation of the services of Wellington and his men. The precious barrels were allowed to settle, bottles and corks were collected from all over the neighbourhood, and in November of that year the men were taken off the farm to help bottle the wine, and to drink the health of the major, the army and the brave new world opening up before them, now that Napoleon's dictatorship was dead. Some thirteen hundred bottles of the best wine were laid in the cellar in racks and bins, and the brown house moth set to work on the corks, for this is another of the substances this ubiquitous insect can digest.

The damage was not noticed for some time for the wine was stacked in the bins, bottle on top of bottle, most of them with their dab of paint to show the upper side, thus enabling the bottle to be brought up without disturbing the deposit. Stacked in this way, the bottles were used from the top downwards and the brown house moth was working the corks from the bottom upwards, in the same way as with the books. When the major's old batman told him he had found a half-empty bottle in the bin the officer took no notice, but by the time three had been found (it was in 1819), the major was convinced that old Tom Sollum was drinking more of his wine than any ex-batman might reasonably be expected to do. A grand investigation was held and old Tom's cry of 'There's maggots in the corks, sir' nearly caused apoplexy in that gallant officer. Hundreds of the bottles had been attacked; where a larva had started to bore inwards a little frass might be seen and where the creature had bored its way out, to pupate, there the little silken case and more frass appeared. As cork is full of cells containing air no problem of the insects' breathing when deep in the cork arose, and if they required moisture they had but to turn to the damp side of the cork, holding back as it was a vast reservoir of liquid. This indeed was their undoing, for the cork was holding back a strongly fortified wine, one containing some eighteen per cent of alcohol, and no animal is likely to live for long once it tries to inhabit such a world. After the excitement had died down, old Tom and the major made a closer examination, and it was seen that the damage was really very little, for any larva that had penetrated far enough to reach the wine was very soon too full of port to continue its activities; in fact most of the insects were repelled by this dangerous fluid. A few larval mines reached so near to the end of the cork that wine seeped into it and slowly penetrated to the outside, which accounted for the three half-empty bottles that had been found, but there were no more than a dozen similarly affected and these were in bottles that had poor quality cork in them, or corks which had already been used once and penetrated by a corkscrew.

'This has got to stop,' said the major, 'and the remedy is wax, hot wax.'

The necks were all cleaned with a rag and the corks carefully painted with a mixture of beeswax and tallow, a task which took Tom and the major a good two weeks. This treatment preserved the corks from further attack. This experiment on the part of the brown house moth was not a very successful one for the race, nor was the major's and Tom's work with the wax very necessary because the wine was slowly seeping into and penetrating the corks, a fractional separation of the water and the alcohol taking place as the alcohol penetrated more rapidly than the water; consequently no moth would have been likely to lay eggs on the cork after the wine had been there for a year, nor after this period would any larva get far in a cork even if eggs had been laid on it, before it was overcome by the fumes. If the major had been bottling a light Bordeaux or some similar wine, the brown house moth would have had better chances of success, because of the much lower alcohol content of such a vintage.

Non-alcoholic fluids would have been still better for the moths, as, in fact, proved the case at Bartons End, for Major Hackshaw had brought back a supply of chalybeate water from the spa of Tunbridge Wells, which he had bottled at the same time as the port, but his life on the farm proved so healthy that he forgot all about it and in the next fifty years all the corks were eaten out by the brown house moth so that the water slowly leaked away.

The spider beetles, hide beetles, carpet beetles or woolly bears also attacked man's goods in the house in much the same way as the clothes moths, but specializing in the high-protein materials such as leather, wool, dried meat, skins, stuffed birds and so forth. The most important family is the *Dermestidae* and nearly all the damage is done by the larvae of the insects, as many of the adults only feed on pollen and nectar from flowers or scarcely feed at all. Like the clothes moths they are adapted to living on foods with low water contents, are great scavengers and would clean up the bodies of dead rodents and birds found in the house. In fact, they are sometimes used to clean a skeleton before it is mounted for lecture and scientific purposes. In the nests of birds and rodents they would live on any high-protein discarded material, such as hair and feathers. Nests and dead animals served as a source of infective material for their attacks on the human stocks of food and clothing. The bacon hanging in the chimney was colonized at times by the bacon beetle, which gave rise to the story that bats eat bacon. Stuffed birds were only another dead animal to these creatures and were soon attacked, but the greatest loss they caused was the destruction of the mask and brush of George Munroyd's fox—a trophy mounted and hung over the fireplace to celebrate a famous run made when he took over the mastership of the pack in 1786. Four years later it had been so eaten by these hide beetles it

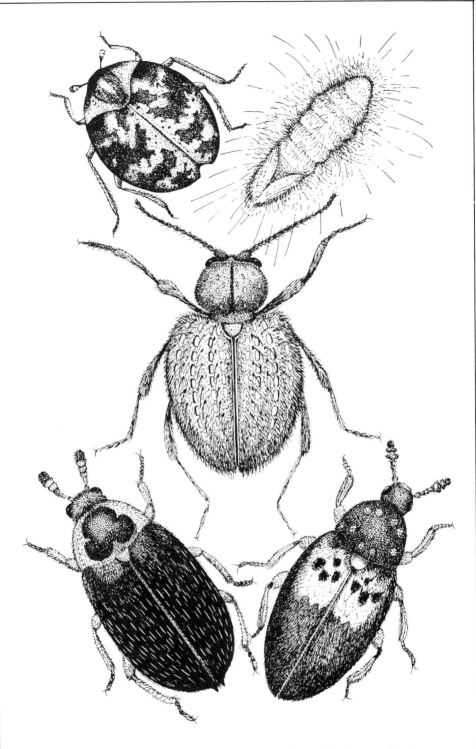

Common carpet beetle and larva (woolly bear), bookworm beetle,
leather beetle and bacon beetle

had to be burned, an event which annoyed the intrepid master very much more than the progress of the French Revolution. Carpets and felting under carpets were also attacked and the woolly-bears are still occasionally found when old felt is taken up, but in general the vacuum cleaner has reduced the population to very small numbers, as it removes eggs, young larvae and adults.

The larvae of these beetles have one habit which causes damage to man. They dig long galleries in inedible substances when they are ready to pupate. They often penetrate the wood of boxes containing food; and they have been known to eat into lead-covered electric cable, which may short-circuit wires or lead to telephone breakdowns. They frequently attack hides and the woodwork of vessels carrying these is particularly liable to be bored into by Dermestids; for instance in 1593 the timbers of the good ship *Thomas Cavendish*, bringing back a cargo of penguin skins, were so used for pupation galleries by these insects moving out of the skins that she took in water and was lost.

The clothes moths were not the only dry-litter insects that benefited from the presence of man at Bartons End. A number of other insects which attacked his food were also found from time to time, particularly in the early days. Some of them showed a considerable adaptation to very specialized environments, and all of them were able to economize their water resources.

In the early days of the farm, the Bartons stored large quantities of grain and flour in the house, as well as the grain stored on the farm, both in bins and unthreshed in the ricks. Regular visits were paid to Richard Barton's water-mill in the village with one or two hundred pounds of grain which were returned as flour a week later and, as the roads might be very bad in winter, a considerable stock of flour was held in the house, at least in the cold months, to provide against the family having to go hungry in bad weather. The miller was John Barton's uncle by marriage, by name Mark Gascoyne; he had married Richard Barton's younger sister, Jane, and had been put into the mill by his brother-in-law, it was said, on account of the size of his thumb. The thumb that went *inside* the measure of every peck of grain, or gallon of flour handled, was thus that much extra profit to the miller, except, of course, when the miller was measuring grain or flour which he took in payment for the work done, instead of money; then he used both hands to grasp the measure, in this way keeping his thumb out of it. The miller's monopoly right to grind corn and the size of the miller's thumb were both constant causes of dispute and jokes of the countryside.

The preservation of the grain sound and wholesome throughout the year was ever an anxiety to the Elizabethans. There were three dangers: the first, that it might not be very healthy to begin with; the second, that it would go mouldy; and the third, that it would be spoilt by insects. Wheat, barley and oats could be attacked by the bunt disease which causes the grain to be filled with a black dust of fungus spores instead of flour. When the attack was bad the release of this black dust into the flour when the grain was ground gave it a fish taste and violet colour that were much disliked; such flour was frequently used for making ginger-bread, when the taste and colour could be disguised.

Wheaten bread was rather a luxury and usually the Bartons used a mixture of wheat, barley and beans. If any of these came too moist from the threshing, or were kept in damp places, they would go mouldy and, in fact, farmers and millers frequently had to dry the grain to prevent this loss—'drying with ashes' is an item often found in millers' accounts.

The third danger was the attack of insects in grain and flour; and the first insect to take advantage of what was for it a large stock of food in the house, was the grain weevil. The weevils are a family of beetles having long blunt snouts which bear antennae about half way along their length, and the grain weevil belongs to the sub-family of weevils that has come to be particularly objectionable to man, for many of them attack his crops, and others the harvest in storage. The apple blossom weevil stops much of the blossom developing, the bean weevils destroy both seed and plant, the cotton boll weevil is notorious, and the turnip gall weevil, palm weevil and rice weevil are a few more.

The grain weevil spread round the world in the train of explorer, trader and conqueror: it has been in England since Roman times. The adult weevil is just under a quarter of an inch in length and is dark brown or black in colour. It is a very strongly protected insect, being well armoured with a coat of thick chitin—the insect integument—and a strong pair of wing cases or elytra, which are possessed by all beetles and are really the forewings. Owing to its sturdy construction, the beetle is able to resist the crushing action of great weights of grain and so survive in conditions which would squash softer insects. When once the beetles get into grain they feed on it, boring in with their long snouts, sometimes consuming the whole of the contents of a grain and at others only taking a taste. After mating the females are ready to lay eggs and show some skill in selecting suitable sites, for the seed chosen for this purpose must not only be big enough to nourish the larvae, but must also contain enough water for the process.

The insects show something of their southern origin in displaying a preference for warm places for depositing their eggs. The female bores out a deep hole in the selected seed with her snout, then turns round and deposits an egg in it, immediately sealing over the hole with a gelatinous excretion which hardens at once. The females prefer grain which offers easy access; the most attacked is hulled barley, after which they prefer rye and then wheat. They do not like unthreshed grain. Up to about five eggs per day may be laid with a total of about two hundred per season. Only one egg is deposited in each seed of small grains such as barley and wheat, but two or three may be laid in such large grains as maize and beans. The adults live about seven or eight months. The eggs hatch in a week to ten days according to the temperature, and the larvae live entirely within the grain, eating out all the interior until only a shell of bran is left; the larva literally creates its own living space by eating away its surroundings, which takes some four to eight weeks, unless a cold period intervenes, when activity ceases, to be resumed with a rise of temperature. The larva pupates within the grain, taking about a week, when the tiny weevil is formed within its shell which it quickly breaks open to reach a more generous supply of oxygen on the exterior. The insect is very vulnerable at this stage as it is soft and can easily be crushed by any movement of the grain. It is at first a light brown in colour and after about four days this darkens as the skin hardens. They are then ready to mate and start the cycle over again; there may be from one to four generations per year. These weevils not only consume a lot of grain, but they foul much more with their excrements, on which bacteria tend to develop, which raises the temperature of the grain store and further hastens the development of the weevils.

In the bad years at Bartons End all this weevil-infested and broken grain had to go to the mill where it made a very poor flour and a worse bread, but in good years with abundant harvests the damaged corn was set aside for feeding poultry and pigs. The Bartons and their successors were well aware of the damage weevils could cause them and the difficulty of assessing the extent of the attack, because the female weevil, in laying the egg, covers her tracks so well that it is almost impossible to see which grain is sound, and which potentially wormy. A rough and ready test had been devised, however, which was to throw a handful of grain into a bowl of water, when the good would sink and the attacked would float.

The Bartons took certain measures against these losses; they knew that grain kept in sealed vessels would remain free from weevil, not only from their own practical experience, but also because it was a measure recommended by the classical authors (it can be found in Pliny's *Natural History*). What happens is that the living grain is respiring, as are also the insects; oxygen is taken in and carbon dioxide given out and, in a sealed vessel, the latter heavy gas accumulates and eventually kills the insects. The Bartons kept a reserve of grain in tightly closed barrels, in stone ewers, and sometimes in open barrels on the top of which a layer of dry sand and road dust was placed. This was successful, provided the grain went into the barrel sufficiently dry, for if it were damp the grain would be free of weevil but would be ruined by mould, or by an attack of grain mites. Grain from the Bartons' harvests usually had a higher water content, than that of North America or Egypt and, consequently, unless the Kentish grain had been exposed to the sun or to the kitchen fire, storing it in barrels beneath a layer of sand was not successful.

Weevils dislike light and air, and the Bartons knew this, for they would take advantage of sunny days to spread out and dry their grain; the movement itself destroys many insects because it breaks open much grain, disturbing the larvae, and also killing many of the soft, newly emerged adults. The drying of the grain, however, was the most important factor; most of it came from the thresher with a water content of fourteen per cent; this the weevil likes and, in fact, can use grain quite easily down to a ten per cent water content, below which the animal is much slower in developing, finding it almost impossible to live in grain with a water content of less than five per cent. If the corn is also treated with an inert sharp dust, such as gypsum or road dust, which is mostly ground-down flint, the dust will absorb water from the grain and also scratch the skin of the beetles and break the protective layer sealing in their water. This has the effect of increasing their need for water, makes them that much more vulnerable and slows or totally destroys their development.

These facts were known in an empirical way to the ancient Egyptians. Joseph, when he collected the surplus harvest of the seven fat years, seems to have stored the grain in the ear in pits beneath a layer of Nile mud which in that climate would rapidly become dust, so that he was able to feed good grain to the population during the seven lean years. In the ancient world people lived close to nature and could solve some problems in an unscientific yet effective manner.

Spilt grain round the farm buildings and in the house is the great reservoir of population for the grain weevil. Little piles of it in corners by the bins, carried down into holes by mice and rats to their winter hoards, and so on, are the places where the insect continues to breed, whatever man may do to keep his main stores free from attack, and where they are still found to this day in spite of the many improvements on the farm. Nevertheless, the population has slowly fallen from the time when the house was first built, because, as civilization developed, less and less whole grain was kept. The Onways, who sold the property in 1689, were the last owners who sent their own grain to the mill and brought it back as flour. The Munroyds purchased flour by the sack from the local miller and baked bread once a week. The Dakers were the last family to bake their own bread as a regular thing, for though Myfanwy Thompson decries modern machine-baked bread as too synthetic for her taste, and occasionally bakes her own, yet when it comes to the point, the labour is too great for home-baking to be done as a routine measure.

All this led to scarcely any grain being kept in the house other than tail corn for chicken feed; the small grain in this is not very acceptable to the weevil both because of its size and the fact that, being small, it dries more readily. There was scarcely any grain weevil in the house during the régime of the Dakers; it was one of the few animals whose numbers grew less at that time. By 1902 there were no grain weevils at all because all the reserves of grain had been used up, mostly by the mice in their efforts to continue their existence in such desirable quarters.

When the Dunchesters bought the place they kept chickens and fed them largely on a purchased grain mixture, as well as on 'laying mash'.

Though this mixture was mostly tail corn it did contain quite a proportion of maize, which was large enough for occasional invasion by the grain weevil, even though it was of poor quality. Consequently this insect is still found at Bartons End from time to time but never in the large numbers common during the first hundred and fifty years of the house. The adult insects either fly in from the nearby farms, are brought in on sacks or are actually in the chicken food when purchased.

When the grain had been converted to flour the material was by no means secure from attack. The grain weevil could still feed on it, though it seldom does, but there are other insects more adapted to living in this loose medium. In nature these insects follow the attack of the grain weevil, consuming the loose flour which falls from corn damaged and broken by the weevil.

The most common followers of the grain weevil are the meal worm, the adults of which are known as the cellar beetle, and the dark meal worm, which are the larvae of two species of beetles, not weevils. The adults of the meal worm are black in colour and they lay eggs (from about eighty to five hundred each) in all kinds of flour, bran, middlings, biscuit meal and so forth, each egg or parcel of eggs being carefully wrapped in a covering of flour kneaded into a paste with saliva to protect it. The larva is known as the yellow meal worm from its shining yellow colour, and when fully grown is about an inch in length, thin and cylindrical with a number of bristle hairs on the body segments, which help it force its way through the flour. This larval stage lasts one and a half years, so that the insect does not breed very quickly compared with the increase in numbers of the grain weevils. It grows fastest in the damper and warmer portions of the medium and can turn to other food as well as flour, as it also eats scraps of meat; this insect has even been found in association with the death watch beetle where it follows up the softening of the wood started by that well-known borer. The dark meal worm has a very similar history; the larva is darker and tinged with brown and its larval stage takes longer—two years. Much smaller though similar grubs were sometimes found in the flour as well, which were the larvae of the common flour beetles.

Little Prudence Onway used to hide a jar of flour at the back of the cupboard in order to encourage meal worms which she used as food for

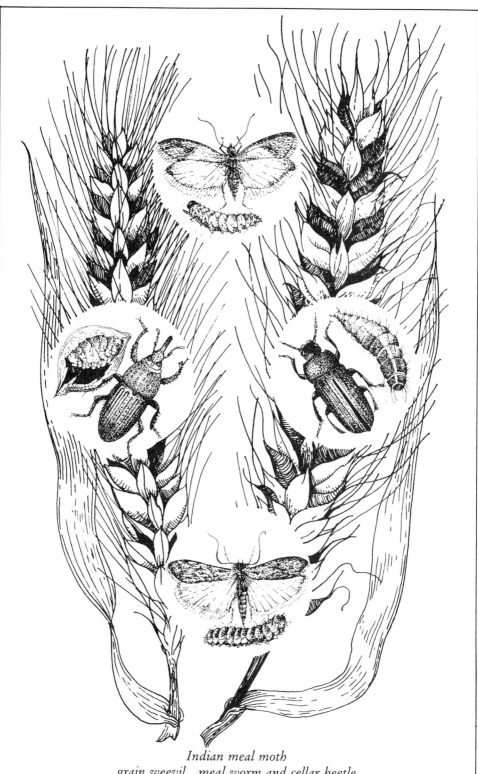

Indian meal moth
grain weevil   meal worm and cellar beetle
Mediterranean flour moth

her animal pets, particularly her tame bat, Master Milton; but she also fed them to robins, sparrows and some toads.

These insects were only found in the house when considerable stocks of flour were kept, or where much grain had been spilt and neglected. Corn stopped going to the mill with the coming of the Munroyds. They still bought flour, a sack or two at a time, and this often contained the eggs, if not young larvae, of the meal worms, although they were not at all common insects in the house, except when specially bred for the purpose of feeding animals. They reached their peak with the Dakers who were careless and spilt grain and flour all over the place, which formed ideal centres of population for this type of dry-litter creature. The insects are not found in the house today because little flour is kept there, and what small supplies there are come from mills which are regularly cleaned and disinfested.

As international commerce increased so did the opportunities for these once rare animals, for wherever man moved his supplies of food there went also the creatures living on it. All the time the men at Bartons End were living on their own resources—on food they themselves raised—they only had to preserve it from attack of the indigenous animals, but when food from other countries started to come in in quantity, new creatures were introduced with it. Many of these died because they could not live under the new conditions, but a few throve because they could and because their natural enemies had been left behind and did not exist in the new environment. Two such animals were the Indian meal moth and the Mediterranean flour moth. The first arrived at Bartons End in the hot summer following the great snowstorm of 1881, in a consignment of flour from a London mill, carrying an infection which had originated in some Manitoba wheat.

The Mediterranean flour moth also came to the house at about the same time, but in French flour, though the original insects came from Central America. Some of this flour also contained the Angoumis grain moth, but it never established itself in the house because the climate was unsuitable: in France it had caused trouble to man's food supplies for many years, particularly in Angoumis from which place it took its name. In the eighteenth century it had caused immense losses there and the British were fortunate that it was never able to adapt itself to the conditions on the other side of the Channel. Were it to be re-introduced to the house today it might thrive because of the artificial climate now maintained there by central heating.

The Indian meal and the Mediterranean moths were never numerous in the house and are not now found there for the same reason.

Chapter 10

Adaptors

THE INSECTS of the dry-litter community were not the only animals at Bartons End which found a prosperous new environment in man's goods; there were many others as well who did not come originally from a dry-litter world but nevertheless adapted themselves to living in or on houses. Some birds and mice are obvious examples, but there are a number of other animals which find life in a house very advantageous though usually in places rather damper than those selected by the true dry-litter animals. Mites and insects were the chief creatures that took advantage of man's providing himself with better food and shelter. The thatch at Bartons End harboured a considerable community of mites at one time, some of them parasitic on other animals, which will be discussed later, but there were three mites which were numerous on food and furniture from time to time.

Mites are not insects; they are very small creatures with eight legs and are related to the spiders; on animals and man they can cause mange and scabies, but at Bartons End the principal mites were those attacking grain and flour. The grain and flour mite followed on the attacks of the grain weevil. They can occur in vast numbers; in grain they bore into the seed and destroy the germ, and in flour they select this part upon which to feed and thus deprive it of a valuable constituent. They give a musty, tainted flavour to cereals when even comparatively small numbers are present; in the early days at Bartons End when complaints were made about the quality of the bread it was not all due to the bunt disease in the wheat, but in part to mites attacking the flour. The flour or grain has to be moderately moist in order for the mite to thrive—they need about twelve per cent of moisture—and a high humidity in the atmosphere; surprisingly enough the mites do not require very high temperatures, being content with a range between 65° to 75°F and as a consequence only certain parts of the flour, such as a damp surface layer, may be attacked.

The mites are very small, they look like dust when they fall from a sack of grain; the adult mites are oval, greyish creatures with eight legs and are about $\frac{1}{50}$ of an inch long. The females lay minute eggs in the flour at a rate of some twenty-five a day, each one some five-thousandths of an inch in diameter. These soon hatch out into six-legged larvae, which at once start to feed and reach their first moult in a few days, after which they emerge with four pairs of legs, feed and grow and moult again; from the second moult they can emerge in one of three forms, either as an adult mite or in one of two kinds of resting stages. In one of these last, the legs are somewhat reduced and the abdomen is provided with a sucker, by which they can attach themselves to other animals frequenting their environment, such as mice, flies, beetles and so forth. In the other resting stage, the legs are much reduced and the body becomes very compact and light so that they can be carried into the air by the wind. In both these states they can endure successfully much more adverse conditions than as eggs, nymphs or adults, for they survive both cold (below freezing-point) and heat (over 100°F); they also resist dry conditions more readily in this form, so that these resting states are a means by which the animal is disseminated.

When these mites became numerous in damp flour, moulds were also found; the movement of the creatures about the flour helped spread the spores of the fungus and so further to spoil the Bartons' hard-won supplies. The flour mites did not have it all their own way though, because they in their turn were preyed upon by other creatures which kept their numbers down. Their biggest enemy was a rather smaller diamond-shaped mite which fed on the flour mite and thus got its food from man at second hand. The predatory mites throve best under warm conditions, whereas the flour mites were more sensitive to humidity, with the result that their relative numbers rose and fell with the seasons.

The flour mite numbers tended to rise in the winter and this rise was followed by an increase of the predatory mites in the summer, due both to the ample supply of food and the increase in temperature.

These mites were quite serious pests of grain and flour in the years of bad harvest weather, because the grain would be damp, as would also the resulting flour. After the beginning of the eighteenth century little grain was kept in the house (though plenty of flour was still stored there), so the mite population began to fall. The decrease was accelerated by the false scorpions, or *Cheliferidae*, which feed both on the actual flour mites themselves and on their predatory associates. These false scorpions are quite small creatures, about ⅛ of an inch long and bear a pair of strong claws very like those of a scorpion. They used to come into the house in the sacks of grain, from whence they would spread to any flour being attacked by mites. They were not very numerous when the only grain actually kept in the house was the chicken food.

The population of mites and their predators in the house again fell when bread was no longer made regularly, as naturally the stocks of flour kept there were much smaller; nevertheless the house has seldom been without them for long. The mites are ubiquitous; their resting stages can be carried on the wind, on the legs of flies, spiders, clothes of man, fur of animals and a host of other ways, so from time to time the cooks and housewives at Bartons End would have the experience of opening the flour bin (particularly if it had not been used for some time) and seeing the surface move gently—a somewhat unnerving experience—due to a massive population of mites. This is encouraged at the present day by the central heating in the house, for provided the flour is sufficiently damp (and a bag may easily get damp if left near a window, from condensation of the hot air on the cold glass) and a few mites reach it by some means, the heat ensures a rapid increase of population.

Another mite found in the house, though not in large numbers, is the cheese mite. It is very similar to the flour mites in appearance, require-ments of moisture, temperature, and life history, except that no resting stage is formed. Small pockets of mites in such cheeses as Stilton add to the 'ripeness'; they are found as small holes with an accumulation of brown dust in them, which dust contains living mites, their cast skins and excreta. In some cheeses, such as the German Altenburger, a colony

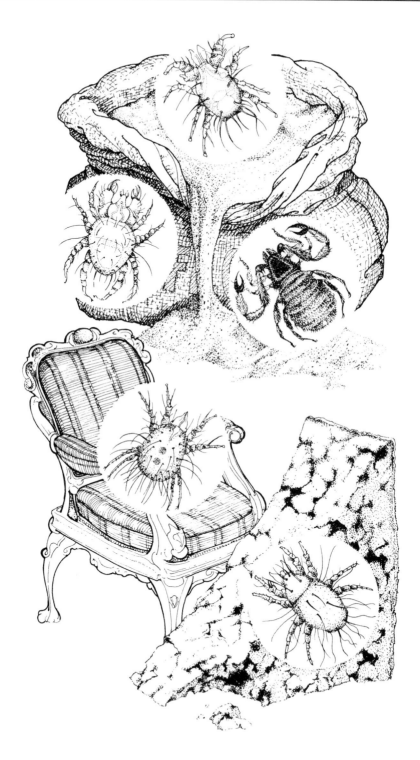

Flour mites and false scorpion, furniture and cheese mites

of mites is deliberately added to the cheese in order to give it its characteristic flavour. Cheese mites must share with oysters and similar shellfish the distinction of being some of the few animal foods still eaten alive by man.

The cheese mite population had two peaks, one during the time of Major Hackshaw and the other during Robert Dunchester's reign. Both were very fond of ripe specimens. Today's inhabitants, the elderly patients now at Bartons End, are not patrons of ripe cheeses. The major used to 'feed' his cheese with port wine, which tended to delay the ripening, for the alcohol at first inhibited the growth of fungi and animals, but as it evaporated it left the cheese pleasantly moist with several cracks and pockets in which mites could develop.

The fourth mite found in any numbers in the house was the furniture mite. It was common in hay and straw on the farm and from here had soon established itself in the thatch where it throve, together with the predators mentioned above. From the thatch it spread at times to the grain and flour in the house and to birds' nests, but its greatest effect on man was when it established itself in the stuffing of William Hackshaw's Victorian furniture. The 'horse-hair' sofa of the drawing-room was actually stuffed with a common substitute for horse-hair—green Algerian fibre—and as the room was seldom used it got very damp, the mite blew in from the farmyard and crept inside the upholstery where it found an ideal environment, feeding on a mould growing on the fibre and the leather. The mites would crawl out of the cracks in the furniture and be seen as a grey dust on the dark leather. Emily Hackshaw often accused her maids of neglecting to dust the parlour, until one day she actually noticed that the dust was moving. She was most concerned and wanted to have the offending article burnt; however her husband had all the leather furniture moved into the barn and sent for the upholsterers, who took out the stuffing, painted all the cracks with turpentine, which killed the mites in them, and some of the wood-boring beetle larvae as well, dressed the leather and replaced the old stuffing with horse-hair; there have been no further outbreaks in furniture, but a hay-box was made during the Second World War, in order to conserve fuel. At the end of the war it lay in a cupboard, damp and forgotten; the moulds started to grow and soon the mites were thriving again—a short-lived triumph which was ended on a bonfire after they had betrayed themselves in the usual way by 'dust' falling from the attacked article.

When the thatch was in bad condition, just before it was removed, being very thick, damp and rotten, this mite was present in enormous numbers, but, of course, they were nearly all destroyed when Joseph Munroyd replaced it with tiles. They continued in odd patches of thatch dropped in the rafters and in damp parts of the cellar where they were found in considerable numbers until 1928, when the whole cellar dried out as the central heating boiler got to work. The mites are, however, still to be found in the house; moulds still grow in damp corners in the roof and in old birds' nests, and winter condensation from time to time provides damp pockets where these and similar creatures establish themselves; they have been known to make a colony in Caroline Thompson's private supply of choice Stilton alongside the true cheese mite.

The thatch also harboured a number of insects known as bristle-tails, and others called spring-tails. The bristle-tails soon found their way into the house, where they throve much more readily than in the thatch. These insects are very primitive, having survived from remote geological time almost unchanged. They normally live in soil, under stones, on rotting wood and so forth but having been introduced to man's environment some of them, in particular two kinds known as the silverfish and the firebrat, find the new circumstances a good field to exploit, one living on sugary and starchy substances, such as the binding of books, and the other exploiting any very warm situation, as under stoves and behind firebacks.

Being primitive insects they have never possessed wings, which makes life simpler for them; insects have no internal skeleton but rely on a strong outer skin to form a support for their organs, having as it were an external skeleton; they grow by casting their outer skins. The old skin is split off several times in the course of an insect's lifetime and each time a new, soft one beneath takes its place, becoming a size larger as it hardens, the creature growing in a series of moults and replacements. Wings, apart from wing-buds, are too complicated to take part in a moult and the change to a fully winged stage is the last moult a to-be-winged insect makes. After it has become winged it can grow no more. The bristle-tails never having had wings, wing-buds or any semblance of such organs, do not have this complication in their life-cycle; they can go on growing by this series of moults for as long as they live, and moreover they can regrow a limb that has been lost or damaged as well; as a result they can vary considerably in size.

The adult silverfish are about half an inch long, shaped like a carrot and with silvery pearl-like scales, which are scattered from time to time as they move around. On the head they bear long antennae and have three large bristles at the rear. They have a pair of compound eyes and somewhat primitive mouth parts. After mating the females lay eggs, either singly or in groups of about three and usually placed in cracks or protected spots, but some eggs are casually dropped as the creatures move about. Each female lays about one hundred eggs, about $\frac{1}{25}$ of an inch long, which soon hatch. They feed on carbohydrates and are particularly active where books and papers are found, scraping away the paper to get at the starch paste which was used in bindings, and actually digesting some of the cellulose. In order to form the substance of their bodies and lay eggs, they have to obtain protein as well which they get from the glue and size in book-bindings and from feeding on the bodies of any dead insects they may find, such as remains cast out by spiders or those creatures whose life spans have come to an end, together with unconsidered trifles such as the empty egg-cases of cockroaches, and so forth. Silverfish shun the light and when uncovered, as for example when a book is removed from a shelf, they stand still for a moment and then run quickly to shelter. Their greatest difficulty in their new medium is to obtain water, consequently they seek situations which are likely to be damp, such as behind loose wallpaper, under sinks, and so forth. The paste on the paper tends to be hygroscopic and remain damp, and not only water is spilt from sinks, but scraps of starchy and sugary food are dropped as well.

Silverfish came into Bartons End from the thatch in the early days of the house; as the books accumulated they throve—the first volume they attacked was Fitzherbert's *Boke of Husbandry* which had been given to John Barton by his father—and they were most numerous there in 1946. In 1947 they were considerably reduced by the insecticidal treatment given to all the woodwork in the house to kill the wood-boring beetles. This did not affect those silverfish actually in books at the time of treatment, nor those behind old wallpaper, and as the insecticide residue on the bookshelves wears off, the silverfish are beginning to build up their numbers again, particularly on the lower and thus damper shelves. Their numbers will decline as the newer books replace the old, for pastes and glues made of synthetic resins, used for book-binding and wall papering, are replacing the old vegetable and animal products, which used to give many insects certain essential elements in their food. Once again a change in the availability of food is more important to the survival of an animal, as a species, than are many of the direct measures taken against it.

The other bristle-tail eventually found at Bartons End was the firebrat, an insect similar to the silverfish and like that creature able to continue to grow as an adult; as many as sixty moultings have been recorded under laboratory conditions. It was not as long-lived as this at Bartons End; it did not thrive in the old kitchen with its open fireplace because there were no warm enough cavities for it, but with the installation of the eighteenth-century kitchen and a number of closed spaces behind the stove, the firebrat found an ideal environment, though eventually it had to compete with the cockroach for it. Here the creature increased its numbers very rapidly, as it can develop in much drier situations than the common silverfish. The firebrats would come out at night and feed on the food scraps left in the kitchen, on shelves, fallen to the floor and so forth. When an electric cooker was installed in the kitchen in 1932, the old kitchen range was no longer used and the firebrat, which so loves the warmth, gradually faded away there. By this time they had already established themselves in the brickwork beneath the central heating boiler in the cellar, a site which suited them well and where again they had to compete with the cockroaches, more success-fully as it turned out because they were more adapted to living on food found in the cellar than the roaches who had to journey to the kitchen for most of their nourishment. Though cockroaches have now been eliminated from the house, the firebrat is still found in small numbers.

Firebrats have a curious courtship procedure, not unlike that of the display of birds. The male dances in circles round the female and repeatedly touches her with his antennae. The actual mating is somewhat akin to that of spiders in that the male deposits a sperm bag in front of the female and then retires, whilst the female herself undertakes the necessary movements to absorb it, if she so desires. There may well be a parallel here between the action of the female firebrat and the

Firebrat

artificial insemination of women, which has caused so much controversy.

The books at Bartons End in addition to serving as food and shelter for silverfish were also occasionally attacked by other insects; there were booklice, bookworms and the clothes moth. The booklice, or *Psocoptera*, are primitive insects related to the true lice found on birds. Those that occurred in the house came from three sources; the first, the thatch, by way of the birds' nests; the second, in a Bible and some other old books brought to the house when John Barton got married, and the third from the natural habitat of these insects—the kindling and logs brought into the house, where they lived under loose bark, in moss and in various damp cracks in the wood. They are small insects, about $\frac{1}{8}$ to $\frac{1}{16}$ of an inch long, yellow or grey in colour and in many cases they are able to dispense with males, the females laying viable eggs without any preliminary fertilization—they are partly parthenogenetic.

The booklice tend to live on moulds and other fungi growing in their habitat, so that a certain degree of dampness is needed before they can attack books. In order to get the necessary protein some of the fungi must be living on substances containing nitrogen, such as the leather or the glue of the bindings. Although the females can reproduce without males, these are sometimes present and can make tapping sounds not unlike those made by the death watch beetle, and serving the same purpose—that of a mating call.

The booklice increased in the house because of the warmth and shelter it provided, though they never became very numerous. During the time it was empty of humans they still continued there, because so much rubbish had been left behind. As the house dried, due to the installation of the central heating, their numbers declined, but they are still there and take advantage of any damp condition to build up the population. For instance, a large colony was found recently living in the sawdust filling of the partition wall between two of the bedrooms in the house, which had become damp from a 'weeping' water pipe, and a lot more are slowly consuming Robert Dunchester's juvenilia still lying forgotten at the bottom of a tin trunk in the loft.

A few books were also penetrated by the bookworm, or bread beetle. This is an insect closely related to the common furniture beetle and was originally a grain-feeder, finding life easier on the debris falling from broken grain rather than in attacking the whole seed. From this field it started living in the stores of food accumulated by hoarding animals, such as mice, rats and squirrels, but soon found that man had bigger and warmer hoards than any animal and so exploited them as far as possible. It has adapted itself to a wide variety of substances, including some strange ones such as belladonna, aconite and strychnine. In bakeries it often attacks bread; of course, this must be left neglected for some weeks if the insect is to complete its cycle of development, so that at Bartons End it bred only in neglected crusts. The eggs were sometimes laid on the old books and the larvae penetrated down into the pages until a full-grown fat grub, turning first to a pupa then to an adult, would burrow out again to the surface to fly around, find a mate and start the cycle again.

When the flying bomb shook the house in 1944, one of the Bartons' old books was discovered in the loft; it was Raleigh's *History of the World* in its original binding, the authenticity of which was vouched for by bookworm burrows coinciding in both the covers and the pages.

Another insect which adapted its ways to living in a house was the house-cricket. It was a creature which was regarded with affection and even encouraged on account of its cheerful song, and it is still sometimes found in the house. The cricket did, in fact, sing for its supper, though now it lives mostly in the cellar and is only occasionally seen in the upper section of the house. At one time crickets would come out at night from behind the big stove in the old kitchen and feed on the scraps of food dropped during the day; they were very numerous around the bread oven and old hearth.

This insect, belonging to the same family as the locusts and grass-hoppers, has its original home in the hot deserts of such regions as Persia and the Sahara, where it finds the climate to its liking. From these areas it has spread in the course of international commerce, concealed in bales of silk, wool, with grain and so forth, but it has also been deliberately carried from place to place by people who enjoyed the sound of its chirruping. The early Chinese much admired the sound and had beautifully carved cages made for their pets. They were popular in Europe too. When Bartons End was built crickets were often given to children as presents. The first ones actually to reach the house were in a consignment of tobacco brought up from the Romney Marsh by

Thomas Barton in 1605, a parcel, incidentally, which had not had the advantage of passing through the hands of the Excise officers, one of a number of such cargoes which added to the prosperity of the family, though not of the country. The crickets spread out of doors in the summer where they enjoyed themselves in the dry grass and cornfields, and sought cover in the autumn. Those that reached a warm house survived, but others that only found a hedge-bottom, or even a barn, were rarely able to endure the winter. Some of them got into rubbish dumps which, if they were fermenting, were warm and thus allowed the animals to live through the cold weather, but in the neighbourhood of Bartons End the race only survived from year to year by colonizing houses for the winter, from which spots they invaded the fields again when the warm weather arrived.

The Bartons and their successors greatly admired the noise of the crickets; young Mark Barton captured one, and kept it in a cage and was very disappointed because it never sang, which was not surprising, because it was a female and only the males chirrup away. Little Prudence Onway, with her great fondness for and understanding of animals, of course, kept a cricket who sang away very energetically and attracted additional females in from the fields. The few crickets present to this very day are regarded by the Thompsons as lucky, but the numbers are much reduced for they had an enemy, not a direct one, but nevertheless a most effective opponent. This was the cockroach, which arrived in the house from the Crimea in 1856, and had the effect of driving out the crickets. It liked the same warm situations and fed on the same food as they did, but was more efficient in exploiting the position, so its numbers increased as the songsters declined. Both creatures disappeared as the house grew cold and empty of humans.

The chirruping of the house-crickets has had a certain survival value for them, for besides attracting the females, the noise has been so appreciated by man that he has tolerated crickets in his buildings and even encouraged them to breed by scattering food and hollowing out spaces beneath hearths for them; not that the noise is now so universally esteemed. Sloane Dunchester disliked it intensely and a cousin of Louise Cavanagh's was said to have committed suicide because of the continued noise of the animals at night. To our ears it sounds something like the squeak of unoiled machinery and may thus be more disturbing to modern man than to his forebears, for our ideal is to live surrounded by well-oiled apparatus which never goes wrong.

The adult female cricket digs a pit (as a locust does) about half an inch deep in soft earth, in which she lays her eggs. The spot chosen must not be too dry or the eggs will shrivel up, or too wet when they will become mouldy. After hatching, the nymphs go through a number of moults, till at the final stage the fully winged insects emerge. The creatures never fly, but the males have an important modification in the wings. The fore pair each have a serrated rasp-like edge, which can be engaged with a hardened ridge on the softer hind wings, so that these two wings rub against each other as they are held aloft at an angle of forty-five degrees, and produce the sound. On a still warm day this can be very powerful and has been known to travel a mile.

In spite of the general favour with which crickets were regarded at Bartons End, they could at times be a nuisance to the humans there, as was found after the hot summer and good harvest of 1802. In the autumn large bands of crickets invaded the house from the fields and started to consume human food wherever they could find it, bit holes in the damp washing (to get water) and almost deafened the Munroyds who swept them up day after day with brooms until cold weather and a food shortage had reduced their numbers to the level of an odd cheerful chirrup from time to time from a few survivors beneath the big old hearth.

Dr Dhusti, the retired Indian scientist and one of the patients at Bartons End, keeps a few crickets in a cage in his room, explaining, to anyone who will listen, that they are sacred animals. The learned doctor has a wry sense of humour and people never quite know whether to believe him or not.

The cockroach was a comparatively recent introduction to Bartons End. This again is an insect of tropical climes which has extended its zone of action by adapting itself to man's buildings, but it was an insect that was slow to spread. When the house was built (1555), the insect was known in London, to which town it had travelled by means of the ships trading with the Mediterranean. It was found in a number of seaports too, and though the goods brought in the ships—the wines, silks, metals, spices and so on—were spread all over the kingdom and some of the roaches must have gone with them, the insects failed to establish themselves in the country. In the late eighteenth century in Hampshire, Gilbert White recorded the cockroach as an unusual insect. However, when the creature does establish itself it is very successful and somehow manages to drive out other insects which want the same shelter and food.

The first cockroaches came to Bartons End from the Crimea, in the spring of 1856 when David Hackshaw's effects were returned to the house, after the death of the young captain in the storming of Sebastopol. As his father, William Hackshaw, was dead and his brother, Alfred, lived in Venice, the War Office returned two boxes of clothes, books and letters to the farm, where Emmanuel Burrows sorrowfully opened them, for he had known young David well when his own father was bailiff on the farm. In the Crimea the camp stores were incredibly dirty and neglected; cockroaches abounded and several were enclosed with the dead captain's belongings before they left Russia. The boxes happened to be carried in a warm part of the hold and the females laid egg pods which were just hatching as the boxes were opened at Bartons End, the young insects being shaken out and soon finding shelter in the old, warm house. Here they prospered, having their chief centre behind the kitchen range and sallying out at night to feed on anything left in the kitchen or larder. Both the Burrows and the Dakers disliked them but were not able to do very much about the matter; they would kill those they saw when coming into a dark room with a lighted candle; but this made very little difference to their numbers.

William Daker, who was always reading books on scientific agri-culture and natural history, was once or twice concerned enough to take some active steps against these insects. He made up poison-baits with bran and Paris green, a mixture which was effective but very dangerous to have in a kitchen as the pigment is an arsenical preparation; fortunately there were no accidents. Another mixture he tried was sugar and borax, which was not very effective though very much safer. He never attempted the real remedy, which was to prevent the cockroaches reaching their warm breeding sites behind the stove, by means of cementing up all cracks and crevices giving access to them.

The roach at Bartons End was the oriental cockroach, or black beetle. They are unpleasant insects and amazingly persistent ones, feeding on all sorts of food, but their main source is human food dropped or left accessible. They will eat any of their own kind who are injured, or unable to escape from a larval skin during a moult, a habit of advantage to the race as it ensures that this valuable scarce food material, particularly the protein, is not wasted. In houses the cockroaches' chief difficulty is water and this becomes acute when a house is centrally heated. It is for this reason that the insects are frequently trapped in basins and sinks—they have been drawn to the water and have been unable to escape, as is also the case with silverfish found in these places. Apart from direct damage done to man's food, cockroaches spoil a lot more by nibbling at it and dropping their unpleasant faeces on it. The badly infected kitchen and the old hall took on the unpleasant smell characteristic of these insects and this was indeed another reason why Oscar Hackshaw eventually changed his mind about living at Bartons End.

It has already been mentioned that the cockroaches drove out most of the crickets and bristle-tails from their quarters as, although very primitive insects, they are efficient ones, but they must have warmth and so when the house became empty in 1899 they quickly declined and were not found after 1901. They were re-introduced to the house in 1920 in a load of coke, but did not become very numerous because the Dunchesters took active steps against them, using large quantities of 'Keating's Powder' for the purpose, a material which consisted largely of pyrethrum at that time. In 1928 central heating was installed throughout the house and the cockroach population began to rise for two reasons: firstly there were now many warm sites where the cockroaches could deposit their egg-masses and, secondly, the pipe runs penetrated all through the house and made it much easier for the insects to reach them, for though they are comparatively large creatures they are also very flat and can squeeze through gaps and cracks in a remarkable fashion.

The owners carried out a campaign against them. One successful move was carefully to cement up all the loose brickwork under the central heating boiler and around the kitchen range. Another, from 1945 onwards, was to use more powerful insecticides such as benzene hexachloride. As a result the cockroach was eliminated by 1948. The anti-woodworm treatment of 1947 had helped. However, the house is not at the moment free from them for they have again been re-

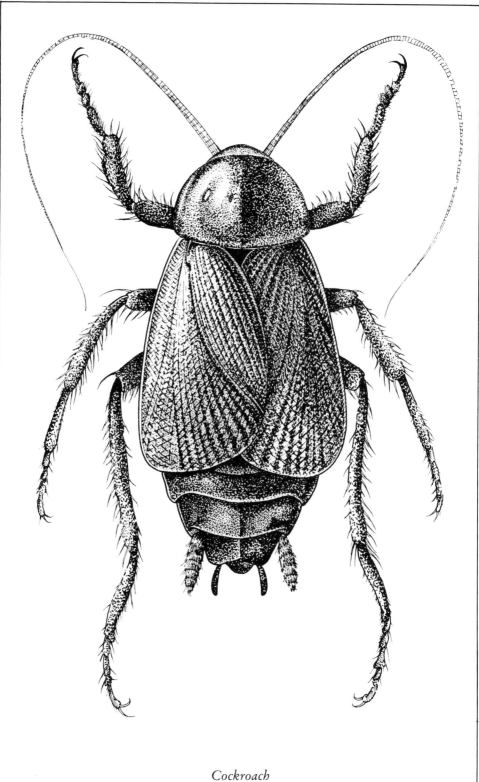

Cockroach

introduced quite recently, this time in the baker's bread basket which was left on the kitchen table whilst the roundsman was having a cup of tea. It is not the same species of insect that caused so much trouble before, but a related one, the 'German' cockroach, or steam-fly. This species differs somewhat from the black beetle or oriental cockroach; the German roach (though Germany is not its country of origin) is smaller and lighter in colour than the black beetle. It can climb better than the black beetle and therefore may be found at any level in a room, instead of only on the floor. The steam-fly, since it can climb, is better able to exploit pipe runs, though it is a thirsty creature and so is more dependent on water than its predecessor, and does not usually climb far from its source of food, whereas the black beetle will make long exploratory journeys.

The German roach is well adapted to living in heated food stores, bakeries and similar industrial premises; it may easily prove a nuisance at Bartons End because the heating pipes there were fitted a long time ago. Where they come through the walls and floors a gap has sometimes been left, and even if the joint was tight and insect-proof when made, expansion and contraction as the pipes heat and cool has opened gaps which will allow the roaches access to a hidden world. However, the Thompsons have many weapons at hand if they wish to fight the insects. They can take out the concealed pipes and leave them exposed, or they can close the gaps with glass wool or cement; they can blow DDT, benzene hexachloride powders and modern insecticides along the pipe runs, they can use insecticidal smokes and sprays, or paint bands of insecticide varnish where the insects run so that it kills them as they cross; this will remain effective for two years at a time. The insects in their turn have some weapons with which to fight the Thompsons, though neither the insects nor, in all probability, the Thompsons, are aware of it at present; these weapons are their rapid rate of breeding and the power of natural selection. Races of roaches may well arise in the house which can tolerate quite large doses of insecticide. It is not likely that an insect which has persisted through millions of generations—from remote geological time—almost unchanged until the present, is going to disappear without a struggle; but still less likely is it that man is going to tolerate the nuisance of such an animal if he does not want to.

A last creature which occasionally was found in the house, taking advantage of man's goods and way of life, was the vinegar eelworm. Eelworms are small free-living worms often found in the soil and in decaying vegetable matter, though some of them, such as the potato eelworm and the sugar-beet eelworm, can cause serious loss of man's crops. In nature, as fruits ripen their sugar content increases and on the skin the 'bloom' develops; this is a wild yeast and, as soon as the fruit cracks with ripeness, the yeast attacks the sugar and multiplies itself, at the same time turning the sugar into alcohol and carbon dioxide gas, which is the basis of wine-making. As the alcohol forms in the pulp, it is in its turn attacked by other bacteria and converted to an acid and water. This reaction is the basis of making vinegar. In the wild state certain eelworms or nematodes thrive in this acetic acid produced by the decaying fruit.

The tiny fruit flies which are attracted to ripe fruit may carry on their legs and bodies the spores of yeast, the vinegar bacteria and the eggs of the vinegar eelworm, and can thus deposit all three to set the process of decay going, converting the complex fruit back to the gas and water from which it originally came; but man has again interfered. From the fruits and cereals he makes wines and beers, and from these vinegars, and if the eelworm finds its way to some vinegar it will multiply very rapidly there, for its food is much more abundant than in the wild state in a rotting fruit.

Vinegar was very important to the Bartons, as it enabled-food to be preserved and relishes prepared to relieve the tedium of the dull winter diet; they regularly brewed it by the old bulk, or Orleans, process, starting from any left-over beer or wine they had. The procedure was to add some liquor, which had started to 'turn' naturally, to a half-empty barrel of beer or wine, keeping it in a warm place, when the vinegar bacteria would start to multiply, at the same time converting the alcohol to the acid. Meanwhile, the liquor was frequently invaded by these worms and a considerable struggle developed in the turning wine between them and the vinegar bacteria. The bacteria require a great deal of oxygen—in fact, they need about half as much in weight of oxygen as there is alcohol in the tub; a pound of alcohol needs half a pound of oxygen to turn it to the vinegar acid, and the bacteria form a gelatinous film over the surface of the liquid. The worms swimming in the mother liquor also need oxygen, like all animals; any oxygen dissolved in the liquid is rapidly used up, so the worms cannot get this much-needed gas from the surface where the bacteria have established their skin of bacterial tissue. The worms struggle to break through this film by repeatedly throwing themselves against it. If the bacteria are active and growing well, the skin on the surface will be too tough for them and the worms will either die and fall to the bottom, or migrate to the sides of the vessel around the orifice, where capillary action raises a little of the liquor and allows the worms to breathe and a small population to survive. If, however, the worms are successful in penetrating the bacterial film when they charge at it, it will be broken up, sinking to the bottom of the vat and will stop the liquor from turning to vinegar. It used to be thought by the Bartons that the process would only be successful if there were plenty of vinegar worms around the bung hole of the fermenting barrel, which in a way was true, though they were confusing cause and effect. The vinegar could have been formed without any worms present, but the presence of worms clustering round the vent hole meant that the bacterial film was strong and whole and the process going well. Today vinegar is made by a similar fermentation process, though a much different procedure is used and vinegar worms are excluded from it. Nevertheless they are still found at times, for the eggs blow about in the air or are still carried by small fruit flies.

Chapter 11

Third Hand

ALL ANIMALS get their living at least at second-hand, for as they cannot themselves directly use the earth, air, water and sunlight to elaborate their food, they must eat plants which can; a number of animals, however (and some plants too), have gone further and get their livelihood at third hand, feeding on living animals which have already built up a supply of suitable carbohydrates and proteins. Such animals are the parasites and predators which were well represented in the house; all the animals there had parasites and most of them, except man, were the victims of predators from time to time as well. The plants which live on other living plants are the parasitic, or disease-producing fungi.

At Bartons End fleas and lice, bugs and certain worms were the earliest parasites of the birds and mammals, while other insects, mostly belonging to the hymenoptera, were parasitic on the insects.

There are different degrees of parasitism and a great number of complexities in the life of parasites. Some, the roundworms, for instance, live entirely within and at the expense of their host: others spend part of their life in one animal and part in another, as for example the sheep liver-fluke which passes part of its life-cycle in a snail. Some parasites are external and can live on several different kinds of animals; others live within the host and specialize in one kind of animal and one situation in it. The effect on the host can vary from no particular harm to actual death; some are only dangerous to the host for a disease they can introduce with their bites, as for instance human lice carrying typhus, or the *Anopheles* mosquitoes malaria; others may confer benefits on the host, in which case the animal is really no longer a parasite, and we give its condition another name—symbiosis. The divergencies and variations in the parasitic pattern are so immense that though we know what we mean by parasitism, we find it a difficult word to define with exactitude.

The life of these animals living in intimate association with other animals falls into three classes of behaviour—symbiosis, commensalism and true parasitism, though the borderlines between them are not clearly defined.

The lichen growing on the wall is an example of symbiosis, as it is an association between an alga and a fungus; both are intimately combined and one cannot thrive without the other.

Commensalism is a term which originally meant the state of two animals sharing the same food ('eating at the same table'), and without doing each other any harm. However, if animals actually eat just the same food, they do each other harm, for as their numbers increase they must come to compete for it sooner or later. Commensalism is better described as the state where two animals live in an association, neither harming the other directly, consequently both can live apart if necessary. The mouse is a commensal with man eating at the same table as he, but in most cases man gets the bigger share of the supplies, so that from the mouse's point of view man is acquiring the status of a dangerous rival. When the situation is reversed, and the mouse starts to dominate the position, then man gets worried and uses his ingenuity to restore his dominance. Another example, where the partners are more beneficial to one another, is that between starlings and sheep, where the birds sitting on the backs of the animals kill the lice and ticks living on the wool and, in the wild state, warn them of the approach of enemies, such as a wolf or a fox, by flying away, usually giving the alarm cry. The sheep benefit from having their parasites destroyed and the birds from a supply of high-protein food, but both sheep and birds can live one without the other.

It is a parasitic state when one member of an animal association definitely harms the other; the parasite may suck in blood or other juices, or if it is an insect, it may live within the animal and allow the host to develop for some time before killing it and emerging as an adult to continue the cycle. Parasitism is a highly specialized state in which sucessive adaptations of the parasite and host usually strike a balance, in which the parasite does not kill the host, or does not kill it quickly, for should it do so it is jeopardizing its own future. It is an indication of a recent parasitic adaptation when a host is rapidly killed. The cabbage-white caterpillar is frequently parasitized by a hymenopterous fly *Apantales*. The young caterpillar is stung by an adult fly, a number of grubs develop and feed within the caterpillar; these grubs allow the host to grow normally but use the food reserves (meant for the caterpillar's metamorphosis to a butterfly) for their own development; in fact, the parasitized caterpillar eats more food than an unattacked one. When the insect seeks a shelter for pupation, and many crept into Bartons End for the purpose, the parasites kill it, consume the reserves, come out from the skin and themselves spin their yellow-white cocoons, from which they shortly emerge as adult flies, to continue the life cycle.

The *Apantales* allows the caterpillar to live for as long as it can be useful to it, taking in this case as full advantage of the situation as is possible and is thus an example of a successful adaptation. This must be the unconscious aim of all parasites, but the matter may become complicated by the fact that the parasite can be the carrier of other organisms, such as microbes or protozoa which cause disease in the host. It is of no advantage to the louse that its human host should die of a typhus infection, a disease which may also be killing the louse, for to continue its existence the louse must find its way to a new host, frequently a matter of some difficulty. Nor is it any advantage to the mosquito that the malaria-causing protozoa weaken or kill men; the malaria organism does not harm the mosquito, but by killing men it does reduce its food supply. These particular insects would both be more successful without the secondary infections they carry.

The field of parasitism is another example of the expansion of life into all those spaces where it can obtain a foothold, even to those corners created by the very expansion itself. To us as humans the word 'parasite' has a base and pejorative meaning, but scientifically parasitism is just another way of life, a way which is at one more remove from the main elements of life, air, earth, water and the sun.

Many kinds of commensals and parasites came to Bartons End from the beginning and quite a number of them have been extinguished there. When John and Mary Barton got married they were free from the blood-suckers such as fleas and lice, but their maid and the old farm labourers soon introduced them. One of the constant tasks of the housewife in those times was to prepare ointments and herb extracts to use against these creatures, with varying degrees of success. It was far easier to keep down the lice than the fleas, for hair would be constantly combed, clothes washed and ointment applied. Lice spend most of their lives on the body (though a fever will drive them off), whereas fleas only attack from time to time and, whilst tending to live on the body when adult, they can hide in bedclothes and cracks and, in any case, lay their eggs and conduct their larval existence quite elsewhere. Flea eggs are laid in crevices in floors and the larvae develop there in dirt and dust. They take but little food but may eat any blood they can find in the excreta of adult fleas or other animals; in due course they pupate and emerge as adults to start the attack again. Fleas have amazing powers of resistance to the absence of food and are extremely sensitive to its arrival. Many people have had the experience of entering a long-empty room and soon being set upon by fleas who have been waiting for a host. The adult fleas are able to remain at rest in their cocoons for a long time and it is the stimulation of the vibration of an animal returning to its burrow, or of a man walking over a floor, that leads them to hatch and, using their powerful legs, hop on to their host, seeking ravenously to feed.

Over the years there were human, cat, dog, bird and rat fleas, and each one preferred to feed on the blood of its own particular host, though it would turn to other animals if its own food could not be found; all were nuisances to their hosts, but the rat fleas were the most noxious to man. Rats became infected with the bubonic plague bacillus during the epidemic and so did their fleas; both died from the plague, but the fleas more slowly, and as the rats died, being short of food, the fleas turned to man, which was how the disease spread. It has already been told how Mark Onway died from the plague, bitten by rat fleas when he had paid a visit to London. Mark's mother, old Elizabeth Onway, still ruled the house and kept everything there spotless—the maids were endlessly employed in beeswaxing the floors though the old lady insisted that elbow-grease was a far more satisfactory substance—so that the plague-infected fleas had no chance to propagate, or to infect fleas on the few mice resident in the house.

Dogs and cats chase fleas in their fur and sometimes catch and swallow them, which is one of the ways in which they can acquire tapeworms; the flea larvae become infected with the bladder-worm stage of the tapeworm as they feed on the dirt in the floor cracks, so that cleanliness is a method of avoiding both disease and worm infections. Rats and mice can also acquire worms in the same way, as could humans, though they are unlikely to swallow fleas. The few humans at Bartons End who became infected with tapeworms did so by eating raw or insufficiently cooked beef or pork, as these animals are the intermediate hosts of the two common human tapeworms.

At one time they were fairly usual among the labourers who would often eat raw pork or beef when preparing slaughtered beasts for the house. They did not do the humans any great damage; it is an extremely ancient parasite in man and has so adapted itself over the centuries as to be singularly harmless; the greatest evil it does, more particularly today, is the shock and horror the individual may suffer on finding he is carrying such a creature. Some few people may be more sensitive to the worm and suffer discomfort, and if the intestinal wall is penetrated this may at times allow undesirable bacteria to enter the blood, in which case the results could be serious. There have been no human tapeworms in the house now for more than a hundred years, due to the decline of home slaughtering and greater care in cooking. Tapeworms are still occasionally found in cats and dogs, when remedial measures are at once put in hand.

To return to the fleas, the biggest population of human fleas built up during the régime of the rather shiftless William Daker when the house was far from clean. It ended abruptly when the place became empty because they were starved out. They were found from time to time during the elder Dunchesters' occupation, particularly during the First World War when shelter was given to a number of war-stricken Belgian refugees who had picked up a considerable colony of fleas in their much harassed flight. They have been virtually extinct since 1922, the main reason being the installation of a vacuum cleaner which clears so much dirt and dust from floors and carpets as to leave no breeding place for the eggs and larvae.

Dog and cat fleas are still occasionally found, but the Dunchesters and Thompsons eliminated these rapidly, the first with D D T, and the second with derris powder, for D D T is very bad for cats. The house martins and sparrows possess their fleas undisturbed by the activities of the humans, and we will discuss these again, and the other bird parasites, later in this chapter.

The three forms of human lice were soon introduced to the house by the Bartons' labourers; they were the body-louse, head-louse and pubic-louse, or crab. Human lice have adapted themselves to living the whole of their lives on or near the body and so enjoy a standardized environment as regards temperature, humidity and food. The body-louse consequently cannot thrive on a naked man; it is an adaptation of the head-louse which came about when man started to wear clothes—an example of the expansion of life into a new field.

Other organisms have taken advantage of lice to introduce to man their forms of life, such as typhus and relapsing fever, and still others get into him when the host scratches because of the irritation of louse bites.

All the humans disliked the presence of lice, though the earlier ones did not regard them with the same disgust as would the present inhabitants; in the early days they were considered unpleasant but more or less inevitable, and one took what steps one could. Pepys recalls that when he, his wife and Deb Willet visited a town near Salisbury they found the beds at the inn 'good but lousey, which made us merry'. One combed one's hair, washed one's clothes and even one's body from time to time and squashed the insects and the 'nits', that is, the grey-white eggs which stuck to hairs or clothing. The body-louse lives mostly in a person's clothes, only visiting the skin to feed when the individual is quiet and at rest. Unlike the head-louse, it can survive a few hours away from the body—for instance when the clothes are taken off at night—but otherwise lice soon succumb to the loss of their chosen environment. Very elementary hygiene serves to keep down body-lice, and as a consequence the head-louse was the commonest of these animals, particularly in children, and was last seen there on some children during the alarm and confusion of 1939, when one authority was billeting families from London to the safety of the Kent countryside, and another was removing children from the danger zone of Kent to the Midlands and west of England.

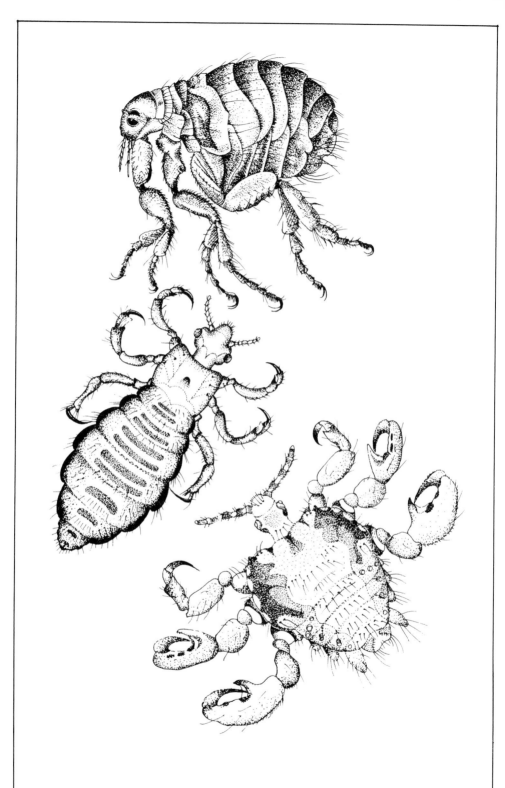

A flea, body-louse and crab-louse

Strangely enough, the wearing of wigs did not increase the infestation of humans with lice and fleas, at least among the prosperous Kentish farmers represented by the Munroyds; in fact, it considerably reduced their numbers. Isaac Munroyd started to wear a wig on Sundays and holidays shortly after he had put a ceiling under the beams of the hall in 1735 and Charles, his son, who married the heiress, continued the practice with greater frequency as he rose in the social scale. These wigs were not only carefully tended, but not worn every day, which meant that any head-lice they picked up would either be killed or die of exposure and starvation before the wig was worn again. Moreover, in order to carry a wig the wearer's own hair had to be kept short, which in itself discouraged lice and fleas, enabling them to be seen and killed by the maid who always dressed her master's hair before he retired. At court and in high society where wigs were worn every day and where at times those of the women were so elaborate and so entwined with their own hair that they would not be undone for weeks, lice and fleas would have much better opportunities, and indeed many ladies had on their dressing tables elegant ivory sticks with a hand or claw at the end which they could push under the wig to scratch their scalps.

The cats, dogs, mice and birds in Bartons End were also infected with various forms of lice from time to time; today the domestic animals are treated with piperonyl butoxide, or derris powder when occasion arises, and so keep down the population.

The next blood-sucker to discuss is the bed-bug which was a comparatively late arrival, for though it was well known in London and the seaports, it did not become general in the countryside until the beginning of the seventeenth century. At Bartons End there were only four brief infections, though a fifth was suspected. They arrived in the house for the first time in some old furniture Mrs Stead sent to her daughter (for the maid's room) shortly after she married Mark Barton in 1604 and were soon eliminated by persistently catching them and painting the beds with sweet oil. The next invasion arose from some of Major Hackshaw's baggage coming from Oporto in 1818 and the third occurred during the very crowded conditions of Emmanuel Burrows's tenancy. Owing to the short duration of these insects' invasions and the dates at which they happened to fall, only the last one appears in the table on page 227.

Bed-bugs are Mediterranean insects which have extended their zone by adapting themselves to the ways of man and birds. They seem originally to have been parasites of bats, and no doubt first adapted themselves to man when both were living together in caves. Bed-bugs need warmth and darkness in which to thrive and, of course, a suitable food supply, which may be one of many warm-blooded animals. Like many creatures which use man, they are much flattened and can creep into cracks, joints, nail holes and suchlike corners in furniture, floors and walls. Often they cluster round the buttons of mattresses; as they find their food in man's beds, they tend to be found in the framework of bedsteads, or in walls near them, and come out to feed on man when he is there, lying quiet, preferring on the whole to do so just before dawn, rather than as soon as it is dark; perhaps the bug has learned that its host is sleeping more soundly then, and it can feed for longer undisturbed, for it needs some five or ten minutes to get a complete meal. The eggs are laid in the same cracks in which the adults pass the day, and are so securely cemented down that the shell remains stuck to its support even after the egg has hatched.

Bug bites cause more irritation and discomfort than those of fleas and lice, but bugs do not seem to transmit any dangerous diseases, as fleas and lice do. Either the opportunity has not yet arisen for some organism to use the bug for this purpose, or if it has, resistant strains of bugs have triumphed over an attempt by an outsider to reduce the food supply. Charles Munroyd made a determined and successful effort to get rid of them in 1750 by using a supply of Mr James Southall's 'Non pareil Liquor' which was advertised as drawing out and killing bugs from their hiding places. Said to have come from Jamaica (perhaps it was an extract of Quassia—an insecticide), it not only made its exploiter prosperous, but also secured him membership of the Royal Society in 1730. In fact, carefully painting all cracks and surfaces of a bed with a feather dipped in turpentine or some oily substance, would destroy most of the bugs present. It is merely a question of doing the job thoroughly.

During the Victorian era, the bed-bug was re-introduced and was continually being fought, for the conditions were advantageous to it. Heavy furniture had plenty of cracks; papered walls provided cover when parts of it came loose; plentiful coal meant the house was much warmer; and large families assured a good supply of food. During this time the presence of bugs carried a certain social stigma; music-hall jokes were made about them and a number of euphemisms were invented for them, such as 'b. flats', or 'mahogany flats' which are descriptive, or 'Norfolk Howards', the origin of which is either the 11th Duke 'distinguished by his habitual slovenliness' or a Mr Joshua Bug, who advertised he was changing his name to Norfolk Howard.

The bed-bug did not have it all its own way, for it was not only bitterly attacked by man but also by another bug, sometimes called the 'assassin bug' from the savageness of its assaults on a large variety of insects and mites. It was also called the 'fly-bug' because it tended to fly into lighted rooms during summer nights; in fact, it still does come into the house in this way; and there are some in the loft preying on insects, mites and spiders. The adult is black with a yellowish tint and in the larval state the insect covers itself with dust, refuse, and cast skins as a disguise, hence its specific name of *'personatus'*, or 'disguised'. Although it attacks and kills bed-bugs it is not above biting man at times and can give a sharp and painful bite which is really worse than that of the bed-bug, but it does not often do this. Assassin bugs are more numerous in the country than in towns and as they attack bed-bugs so readily, they helped to ensure that there were fewer bugs in the country than in the town. Though in this country mostly beneficial to man, similar species in South America can not only cause painful bites but transmit serious diseases similar to malaria, such as 'chagas disease'.

Very soon after the house fell empty in 1899 it became free of bed-bugs, for besides being attacked by the assassins they were unable to last out even an averagely cold winter; they have not entered it again, although there was one unfortunate incident recently. In 1957, in the early spring, the Dunchesters acquired a maid from Austria just at the time the outside of the house was being painted. A few days after the arrival of Anna Dopfler, they were horrified to find bed-bugs in their bedroom and could only conclude that the girl had brought them with her. A series of misunderstandings conducted in halting German and English reduced Anna to tears and filled the Dunchesters with despair, until at last they consulted the distinguished entomologist of the local research station as to the course of action they should take in face of this threat. Somewhat sardonically he said that all they could do was apologize and perhaps improve their German, for the bugs were not bed-bugs but the swallow-bug from the house martin nests, which had taken refuge in the house because the painters were knocking them down. The two insects were very similar and there is yet a third kind — the pigeon bug.

The presence of these bugs in house martins' nests is one of the reasons why it is no unkindness to the birds to destroy the nests at the end of the season, though perhaps it would be better still to sprinkle them with insecticide powder and thus leave the returning birds the same, but pest-free home. However, this course of action might only have the effect of encouraging sparrows, who so frequently take possession of the house martins' nests in winter and who would thrive just that little bit more were they free from parasites.

Fleas, lice and mites are also found in the nests, the fleas being comparatively recent transfers from mammalian hosts and the lice feeding mostly on the feathers. Birds can harbour such a big community of life that A. E. Shipley said, 'Birds are not only birds but aviating zoological gardens.' The larvae of bird fleas need comparatively humid conditions, consequently they are not much found in high, airy, dry nests but in those made of mud, or near the ground, and the house martins' nests provide splendid opportunities for them. The adult fleas spend most of their time in the nest itself and only go on the bird to feed, though they may take immense quantities of blood when they do so, most of which is excreted unchanged into the nest and serves as food for the larvae.

Fleas are not as closely bound to a host as lice and, though free, have few powers of searching for a bird, finding a new host more by luck than judgement; however, they are assisted in this by being able to do without food for a very long time.

The enormously strong legs of fleas serve, in place of the wings they have lost, to enable them to hop away from enemies and find new hosts. They are attracted by warmth and are sensitive to smell, and the closely related sense of taste, and each of the various fleas shows a marked preference for one host, though they may be able to feed on several. Horses have been able to develop (or perhaps it is a chance adaptation) a smell which repels fleas and are not attacked by them, nor are grooms, provided they do not wash too much, one of the few advantages of uncleanliness.

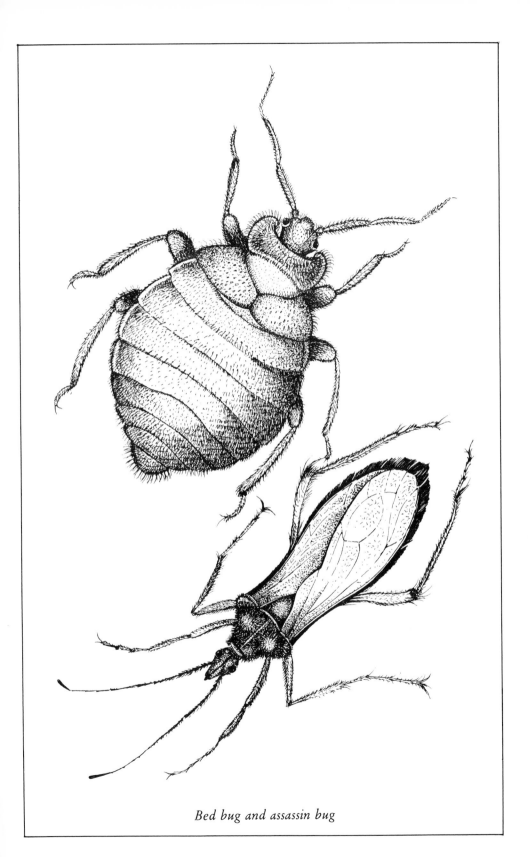

Bed bug and assassin bug

House martins suffer very much from fleas because they continually return to the same nests, for hundreds may be found at a single site; if a nest is abandoned, the fleas migrate and some find their way into the house, where they will slowly perish as they vainly search for more house martins; frequently they may alarm the humans there in the process.

The fleas have their enemies; man, of course, is one of them. He attempts to destroy his fleas as best he can and has done so from early times; one recalls De la Tour's marvellous picture, painted in the first half of the seventeenth century, of the woman, lit by a solitary bright candle, squashing a flea between her thumbnails. Similarly birds, in preening themselves, destroy large numbers of fleas and desiccate many more by taking dust baths. It is doubtful if splashing in water rids the animal of fleas; it may even help the insects as they like a humid atmosphere.

The rat flea can get the plague and die of it, the gut being blocked by the growth of the bacilli and the flea starving to death. The insect infects man with the disease because it cannot pass the blood through its system as is normal, but has to regurgitate it through its mouth mixed with the bacteria, which then gain access to the man through a fleabite, and set up the disease. Bird fleas can be attacked by protozoa and at times half or more of a flea population may be infested. They are also attacked by roundworms which, feeding on the generative organs, have the effect of sterilizing them. Other worms use fleas as the intermediate host. Certain beetles devour fleas in large numbers as do also ants if they can get access to the birds' nests.

The most serious enemies of the nest-inhabiting fleas are the mites, several species of which live in the nests and destroy larvae and pupae, making one wonder whether Dean Swift's* famous stanza arose from observing fleas with a magnifying glass, or poets with his yellow eye,

* So naturalists observe, a flea
  Hath smaller fleas that on him prey;
  And these have smaller fleas to bite 'em,
  And so proceed *ad infinitum*
  Thus every poet, in his kind
  Is bit by him that comes behind.
                              Pastoral Dialogues

leaving an observation which was triumphantly verified when the microscope showed him to be right.

These flea-destroying mites, when in the resting stage, attach themselves to fleas, nearly always choosing the females, as only these will have larvae, and are transported to new nests.

When a bird dies the fleas leave it and seek a new host, which, of course, they may be unable to find, and when a flea dies the mites, both in the resting stage and otherwise, leave it to do likewise. Birds frequently preen themselves, so the fleas have to live in the nest most of the time to escape these attentions. However, they are not safe here for all these other dangers await them.

The bird lice have chewing mouth-parts and so bite and masticate their food, which is mostly feathers, in contrast to the fleas who suck blood from their hosts. Lice are far more successful on birds than on mammals; there are hundreds of species that attack birds, and relatively few that live on mammals. These mallophaga (or 'wool-eaters') must first have been scavengers who adapted themselves to feeding on the reptiles which were the predecessors of birds and are consequently a very ancient order of insects, being but little changed over the evolutionary centuries. On the other hand, birds have greatly changed, and it is sometimes possible for scientists to discover to which families birds belong by making a study of their feather-lice, for it is most likely that birds were parasitized at an early stage in their evolution. All the game-birds have the same kinds of lice, all the ducks and so on, and whilst the birds have changed, the lice have not. The feather-louse lives most of its life on the bird in contrast to the flea which only visits the bird to feed. The eggs are cemented to the feathers, usually along the barbs against the stem or rachis of the flight feathers, or on those of the head and neck. They soon hatch but in a curious manner; the young louse within its shell sucks in air which is blown out behind itself and then bursts the eggshell open. The lice have no metamorphosis, as do butterflies and so forth, but grow and moult, grow and moult three times before becoming adult. They live by cutting off pieces of feather and chewing them, though some species take blood as well as skin, scales and serum, and have thus adapted themselves to a difficult diet.

Unless their population is very high the lice do little harm to the birds; on the small passerine birds, such as our sparrows, the numbers of lice are few, from none to ten in number is common, but two thousand can often be found on large birds. Heavy populations prove excessively irritating to birds; even Pliny remarks that lice can kill pheasants unless they rid themselves of the parasites by means of dust baths.

As the lice spend their whole time on the bird, they are living almost under incubator conditions and in these circumstances no very large number of eggs is needed to secure the survival of the race. They are very sensitive to temperature, and if a bird dies soon leave the body to seek a new host, a matter in which they cannot very often be successful. One can imagine it as the penalty of their comparatively easy parasitic life. It is the same problem that faces the human parasite—the great man's hanger-on. What does he do when his patron dies, or falls from power? Usually he must, like the louse, fall from power, too.

Yet the race does survive and so individuals must pass from one bird to another, for which purpose a number of different methods are used; there is contact between birds during copulation, brooding of the young and roosting. Lice adhere to bird-flies and so are transported, and even to other insects such as fleas, mosquitoes, dragon-flies, bees and butterflies; as flies tend to visit a bird soon after it is dead they find a warm atmosphere prevailing, that is when the fly is on a warm-blooded animal. However, any particular feather-louse is usually only adapted to feeding on one kind of bird, or just a few kinds, so that it will not be successful unless the fly transports it fairly rapidly to a similar kind of bird, an event which cannot be common. Whilst on the fly or other insect, the louse rapidly cools and becomes torpid, a condition in which it will not long survive. In view of all these difficulties it is surprising that there are so many species of bird lice; one would imagine it would be so much easier for the house martin louse to be able to live on any other bird rather than only on house martins, but the fact is that isolation tends to breed species, and the lice on their specific bird are rather like the animals and plants on an isolated island, where all sorts of strange, uncommon varieties can arise.

Apart from the birds themselves, the lice do not have many enemies, but nevertheless they do have their difficulties, as we have just seen, as do all varieties of life in their struggle for existence.

In preening themselves birds not only pick off fleas but also clean up lice where they can, to which action lice have responded by acquiring protective coloration, or siting themselves where the beak cannot reach them, on the head and neck. In many cases lice in these last sites do not trouble about a camouflage colour, although as birds sometimes help their mates by cleaning each other's head and neck feathers, some protective colouring here could be of use. They shake off lice in dust and water baths, though they may also pick them up in a dust bath if another and lousy bird has just used it. Finally, birds may have another strange anti-louse measure in 'anting'. This is when they allow a swarm of ants to run through their feathers and over their skins. In other cases birds will pick up an ant and rapidly dab the insect under first one wing, then the other, then drop it, usually with the abdomen broken, and repeat the process several times; it is more than likely that the formic acid from the ants either repels or kills the lice. Birds will 'ant' with other substances as well, such as hot charcoal from bonfires and even burning matches. Maurice Burton thinks that this may well be the origin of the story of the Phoenix rising from the ashes, for a bird rising from a bonfire site might well be thought to have been born in it.

A great variety of other creatures is found in the house martins' and house sparrows' nests on the walls, eaves and roof of the house. In fact, the nest fauna is almost as great and as varied as that of Bartons End itself, so that we can only consider some its members. There is a big range of parasitic worms, with their complicated life histories. In vertebrates, particularly warm-blooded ones, worms find the best conditions: the medium is larger, standardized and usually warmer than is that of an insect, and the vertebrates moreover wander over a bigger area of ground allowing the eggs to scatter more widely and with a better chance of finding the primary host. Were man an insect-eating creature his worm population would be as high as that of the insect-eating birds and the bats. This subject will be referred to again later in this chapter.

Louse flies and black flies are important inhabitants of the nests; about half the population of birds can be carrying these insects. It is suspected that the louse flies carry the avian malaria parasite—the creature with which Sir Ronald Ross conducted his epoch-making experiments in India in 1897 and thus solved the human malaria problem.

The adult louse flies live permanently on the bodies of the house martins and other birds and run in and out of the feathers in the same way as does the bat fly described in Chapter 6, to which they are related. The insects suck the blood of birds and will also bite man and other mammals. As with all parasites living permanently on a host, large numbers of eggs or larvae are not needed to secure the future of the species. The larvae of the bird louse flies are, in fact, raised one at a time from an egg retained in the body of the female and incubated there, the larva being deposited fully grown in the bottom of a nest, where it immediately pupates, in which state it passes the winter, hatching in the subsequent spring. The adults in their protected environment live quite a long while and have almost lost their wings.

Mosquitoes and gnats feed on birds and are dealt with in the next chapter as only part of their life is spent in the house, the larval stage taking place in water. Black flies are mostly pests of mammals but some attack birds; here again, the larvae are aquatic and are found in swiftly-running streams clinging to rocks, stones and submerged branches. The adults have a curious way of emerging from the cocoon; by accumulating a pressure of air within the pupal case the adults burst it open, rise to the surface with the emerging bubble of air and immediately fly away. It is a matter of some difficulty to transfer from one medium to another because of the small size of the creatures involved; few insects can drink from the surface of water because the surface tension and capillary action of the water can grasp them and hold them there. If the black fly does not at once fly away as its bubble bursts, it will be caught by the tension of the surface skin of the water and be most unlikely ever to escape from it.

The horse-biting stable-flies sometimes attack birds, and bluebottles and greenbottles breed in their dead bodies. The special birdbottle fly lives on the nectar of flowers as an adult but the larvae live in the nest and at intervals suck blood from the nestlings. There are a number of other flies, as well as those mentioned, and all help spread the diseases to which birds are heir.

A large number of different kinds of mite is found on birds and in their nests. The red mites suck blood, others eat feathers; some are scavengers and yet others feed on the parasitic mites themselves.

Ticks can be very dangerous to a bird. They tend to attack it in the region of the eye and, as they gorge themselves on blood, they inject saliva into it which may blind the unfortunate creature. Amongst these animals living on the birds are others which live on the parasites, such as the predatory mites and histerid beetles already mentioned, fierce anthocorid bugs who are greedy for mites, and many others involving a whole complex pattern of life cycles dependent at third and fourth hand on the food arising originally in the green plant.

Of the large number of animals found in the house martins' nests fleas predominated, next came flies, then moths, among these last being the common clothes moth and related insects. It is indeed strange that these birds can apparently be so cheerful and energetic when they have this big colony of parasitic life dependent on them; one can only suppose that they get their food comparatively easily and that the parasites somehow regulate their appetites so as not to kill the goose which lays the golden eggs.

We must now return to some of the other third-handers found on the premises, and will start with that complex subject—the parasitic worms of the mammals.

The parasitic mode of life of these worms has no doubt arisen in the course of evolution from a free-living life and it is tempting to imagine that the creature has given up its freedom for a life of ease—no question here of 'give me liberty or give me death', a statement which does not arise in nature. Actually it is merely life once again filling up a gap—taking advantage of space and opportunities that are not being used. In many cases the life pattern of parasitic worms is so complex that one is tempted to imagine that it cannot have arisen by means of natural selection, though one would be wrong. As an example the eggs of the fish tapeworm are laid in the small intestine of man, passing out in excreta, after which they must find their way to fresh water to hatch and swim. Many die, but some will be eaten by a small crustacean which in turn must be consumed by a trout, salmon, perch or eel and then eaten raw or undercooked by man in order to complete its life-cycle, which can be completed in no other way. Similarly, a cat liver-fluke uses a snail and a fish as intermediate hosts. Have these complex patterns arisen by evolution, by chance or by Divine intervention? Strange as these life-histories are, evolution—the survival of the fittest to survive—can explain them. Our surprise at the existence of successive hosts arises from considering ourselves the centre of things, and working outwards from this point. How surprising a story it seems that man should pass the eggs first to the crustacean cyclops, then to the fish, before they get back to man, whereas the reverse process could be the evolutionary pattern. The addition of successive animals to a simple parasitic relationship gives the parasite a bigger field and a wider range. Some of the worms in the cyclops hatched when their host was eaten by fish and found they could live there with advantage, so they throve at the expense of those that did not hatch inside fish; again some of those in the fish when eaten by man found this advantageous to the race as a whole, for birds and mammals are able to scatter the eggs over a far wider field than fishes and crustaceans, and soon this pattern was obligatory for the species. Such a design is not, of course, fixed for ever. Evolution is not something that has happened in the past and has then stopped, but is going on all around us all the time. Parasitic worms are now rare in man and are being rapidly reduced in our domestic animals, because man cooks most of his food and takes measures against the worms in his animals, destroying the intermediate hosts, using drugs to kill the worms and maintaining a high standard of general cleanliness. This has left an empty field and we may well ask 'Can nature fill it?'

The most dangerous worm found in man at Bartons End was the *Trichina*. The life-cycle here can be carried out wholly in one animal and pigs, humans and rats are commonly infected. No eggs are excreted but the worm forms cysts in the muscle flesh of its host which hatch when this flesh is eaten by any other suitable mammal. The worms can set up a painful condition known as trichinosis, when severe rheumatic pains arise wherever the larvae are turning to cysts in the muscles. Pigs usually become infected by eating rats, and man by eating raw pig meat, or insufficiently cooked pork. Infected pork or ham is excessively dangerous; the tiny cysts have a hard shell and can be felt against the knife edge as its cuts a slice. The worm is rare in man today because of meat inspection and careful cooking, but there is still a danger, for inspection may miss an infection and cooking must be long and thorough to kill deep-seated cysts; for instance, all parts of the joint must be raised to a temperature of 137°F.

The dogs in the house were atacked from time to time by a number of worms of which the dog roundworms and three tapeworms were the most usual. Many dogs, if not most of them, are born infected with roundworms. Eggs are passed out with excreta and do not become

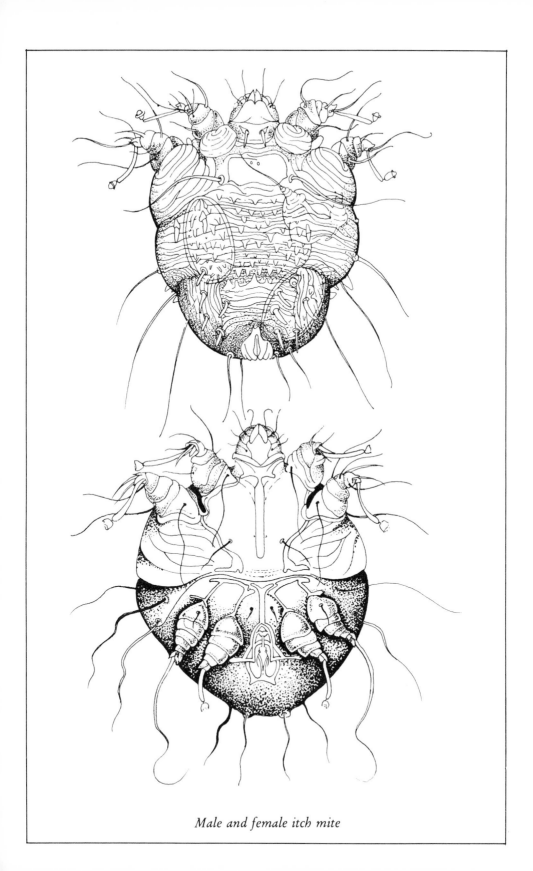

Male and female itch mite

infective for several days; when eggs in the right stage are swallowed the embryo worm hatches in the intestine, bores through its wall and gets into the bloodstream and from there to the lungs, where it can cause a type of pneumonia; from here it passes to the small intestine, where it becomes mature and produces eggs again. Provided they are not too numerous the worms do little or no harm to the dog in this position, though they can inhibit growth and harm puppies.

The secondary hosts of the two dog tapeworms are fleas and rabbits, dogs becoming infected either by eating fleas or raw rabbit containing the appropriate stages of the worms. Sheep can also form a source of infection for other worms. Today the most usual worm-killing drugs are piperazine with castor oil for roundworms and arecoline-acetarsol for tapeworms. The droppings from infected dogs contain millions of eggs and should be collected and burnt.

The small hydatid tapeworm can attack dogs, cats and man, the last-named being infected from the handling of dogs, from being licked by them, or eating raw vegetables fouled by dogs; it will not reach sexual maturity in the cat.

Cats may also be attacked by roundworms and tapeworms, the dog-flea tapeworm being very common in cats as well.

I have mentioned mites attacking birds or their parasites, and living on debris or food in the house; they can also live on the mammals there, when the results are known as mange and scabies.

Mange in the dogs was usually caused by the sarcoptic mites which live among the hair and on the surface of the skin. The young mites burrow into the upper layer of the skin over which point a scab forms from the yellowish fluid seeping from the entry hole. The scab protects the mites who then lay eggs and extend the scabby condition of the animal, whilst the hair falls away. Scabies in man is caused by a variety of the same mite and is a very ancient trouble to him. The itch was well known in the early days of the farm. The female mite burrows into the surface of the skin, is fertilized by the male, and then continues her burrow, laying eggs as she goes. The mite carries a number of backward-pointing spines so she is never able to retreat from her burrow and after something like eight weeks she dies at the end of it, having laid some two eggs daily during the period.

These activities produce scabs on the skin which are very irritating, causing the patient to scratch himself violently and open the skin to secondary infections. The curious thing is that the irritation does not necessarily occur when the mites are working; the attacked individuals become very sensitive to the mite, for whereas the first time a person harbours these creatures the itch will not start for three or four weeks, after he has become sensitized by one attack he will find himself wanting to scratch within a day or two of a new one starting. The hands and wrists are the areas most frequently attacked and the head but rarely. The young stages of the mite have a high mortality rate and the parasite is transmitted from person to person by contact of hands or bodies, with the passage of the adult female mite.

Another form of mange was also found at times on the dogs and was a much more serious condition for them; this was the follicular or demodectic mange, which like the sarcoptic is also caused by a mite. This time it is a long worm-shaped creature, only $\frac{1}{80}$ of an inch in length, unlike the sarcoptic mange mite which is round. The mites penetrate the tubes from which the hairs spring—the famous follicles of the barber giving a sales talk on hair restorer. These mites can also attack man but do him no particular harm, although on dogs they can cause a severe and unpleasant mange.

Similar mites cause the mange of sheep which has been overcome now for more than a hundred years by the invention of sheep dips based on sulphur or arsenic. Today better and less toxic chemicals are used, such as diazinon, one of the organo-phosphorus compounds.

Bartons End is to some extent a microcosm of the world in general and all the animals it has sheltered have reacted one on the other all the time, but it is these parasitic third-handers that have been most influenced by the development of the house, for nearly all of them have been extinguished from the premises and from much of the world outside as well. While there are places, particularly in the tropics, where fleas, lice and worms make man miserable, there are many cities where it is almost impossible to find these creatures. They have been abolished and have left an unfilled space; or is it unfilled? We might well say that it is man himself who has filled the gap. He can do with less food because he does not have to feed his parasites as well as himself, and doing with less food he does not, in this country, have to draw so much from abroad, leaving there a balance to enable his own numbers to increase.

Has man in destroying his parasite lost anything of advantage? One would think not, but parasitism is a strange relationship and can border on symbiosis, so that one can ask questions which cannot really be answered. Did the tapeworms supply a vitamin or a protective antibody which helped man to survive? Did fleas and lice form pets or companions for oppressed peasants or labourers? The Eskimos tell fairy stories in which lice are the hero and heroine, and with regard to dogs, David Harum thought '. . . a reasonable amount of fleas is good for a dog—keeps him from broodin' over being a dog, mebbe!' Were these parasites advantageous in that the misery they caused led man to his seeking to make a better world?

The tolerance some animals show for their parasites suggests that there is something about their activities that the host enjoys, even though it is being harmed. The bat, an excessively clean animal, takes no notice of the large bat flies in its fur, and may get some pleasure from their movements; it is possibly analogous to the pleasure man gets from tobacco, which he continues to use though he knows it disposes him to lung cancer. It may well happen that an animal will not always do what is the absolute best for itself, but can be influenced by the activity of some other form of life to adopt a slightly lower standard and thus permit this other form of life to expand too. Certainly if the tobacco plant had not developed its particular alkaloid—nicotine—so attractive to man, it would be an insignificant American weed instead of, as it is, a major crop plant covering thousands of acres throughout the world.

Something of the same thing may happen with the cuckoo, which parasitizes many kinds of birds, some of whom realize the cuckoo is an enemy and attempt to drive off the invader. She never fights back for she would not want to injure her offspring's foster parents, so she retires and waits for another opportunity to lay her egg. Other birds seem to welcome the cuckoo and almost invite her to lay in the nest, though as a result their own chicks will be destroyed. Moreover, foster parents, such as the small meadow-pippits, have to work tremendously hard to satisfy the enormous cuckoo in the nest; is it possible that they, too, get some psychological satisfaction from these activities?

🪰 *Chapter 12* 🪰

Visitors

P TO the present I have only been writing about animals (at any rate the principal ones) that live, or have lived, permanently at Bartons End, using the place as their home, though some of them might range far from it, as for instance the house martin and the fish tapeworm. There were many such permanent residents, ranging from mite to mammal: a similar range of creatures visited the house from time to time and I will now give an account of some of the more important of these visitors.

Some came regularly at definite times of the year—for example the house-flies—whilst others were casual arrivals who might visit at any moment—an example of this last being the humans and rats, though even these tended to come more during the summer than the winter. Not all the animal visitors were summer ones, for many creatures chose the house as their winter quarters. Let us consider first the summer animals, of which the house-flies were the commonest.

At first sight it might seem that a house is a somewhat abnormal environment for flies, for the larvae live in excreta and decaying organic matter on which the adults also feed. However, some flies have adapted themselves to feeding on man's food as adults as well as on his faeces and body fluids, such as sweat and the mucous liquid of the eyes and nose; yet others want his blood.

The commonest flies were the house-fly, the lesser house-fly, the stable-fly and the fruit or vinegar-fly. There were occasional invasions of the cluster-flies as well.

The two house-flies bred in the dung in the farmyard, preferring horse, human and pig dung in that order. They do not like cow dung. Such dung heaps, due to bacterial action, are usually hot, which makes them attractive to flies. Decaying and heating piles of organic refuse, such as compost-heaps, were also used for egg-laying. The females laid the eggs in cracks in the dung heaps and the blind, almost headless and rather revolting maggots hatching from them bored their way into the interior of the heap, seeking a zone with a comparatively high temperature; they liked one of about 115°F. After three moults the larvae sought a place to pupate, usually digging well into the soil for this purpose and choosing a far cooler situation. They would even leave the heap if it was too hot for them and travel some yards over the ground. The complicated transformation from a legless, blind maggot to an elaborate, winged, keen-sighted fly took place inside the pupal case, the end of which was burst open, when the time came, by the adult alternately blowing up and deflating a special bag carried between the eyes known as the ptilinum; it is the only time this organ is used. When the adult emerged it had to pause for some while to pump its wings full of blood and dry and harden them before it was ready to fly away, feed and renew the cycle.

The common house-fly was that most seen, but during the cooler weather the lesser house-fly also appeared: the stable-fly was a frequent summer visitor, giving rise to the belief that house-flies could sting, for the stable-fly sucks blood and is not unlike a house-fly in appearance. The two house-flies, feeding indiscriminately on excreta and human food, can transmit diseases which occur on the former material and they were in the past one of the causes of the high infant death-rate, for they were the usual agent of the transmission of enteric fever and diarrhoea, which helped kill so many of the babies. Today there are not only fewer flies but children and their food are better protected from these insects.

The humans always found the flies annoying, but in trying to get rid of them usually attempted the impossible by attacking the flies themselves rather than the larvae. The chief characteristic of flies is that they fly; another is their rate of rapid breeding so that killing a roomful of these strong-flying, fertile creatures does not do much to lower the number of flies coming forward to invade the room, attracted to the house by the smell of humans, food and a lower light intensity.

A succession of measures was used against flies, few with much success until quite recently. In the early days, even in summer the maids were up before sunrise. Before retiring at night the curtains would be drawn back and the shutters opened on an eastern window which led the flies to accumulate there; the first task of the girls was to get rid of these by killing them with a fly-swatter or by shooing them out of the window. Another method of keeping a room free of flies—as effective today as then—was to kill all the flies in it and then to keep the doors and windows shut. This was all right for the parlour, but not of much use in the kitchen, where the nuisance was greatest.

In the eighteenth century fly-baits were used, consisting of decoctions of fleabane, wormwood and other plants, with a little sugar or honey as an attractant, but they were not very effective, nor, had they been, would it have made much difference to the general fly population. Major Hackshaw brought back anti-fly measures from the Peninsular, where the creatures were a far greater pest than at home. One was the sealed and darkened room, which was much cooler in the heat of summer than one with the windows open, and another was a bead and bamboo curtain over the doors, for though flies could easily pass between the hanging strings of beads they very much hesitate to do so, and the device enabled the major to snooze undisturbed through many a hot afternoon of summer.

In the mid-nineteenth century fly-papers were invented; the first were of absorbent paper containing arsenic and sugar, which were dampened and left exposed for flies. They were very effective but dangerous as the arsenic was poisonous to all animals. Many accidents to children occurred and even some murders, so the system was soon abandoned. The next measure was the sticky fly-paper hung from the lamp in the middle of the room. This was apt to clear off the lesser house-fly more rapidly than the common fly, as the lesser tends to circle high in the centre of the room. About this time a general understanding of the nature of the problem arose and it was at once seen that the real solution of the fly menace was to stop the larvae developing. The house-flies do not breed much in loose droppings but mostly in dung and compost-heaps.

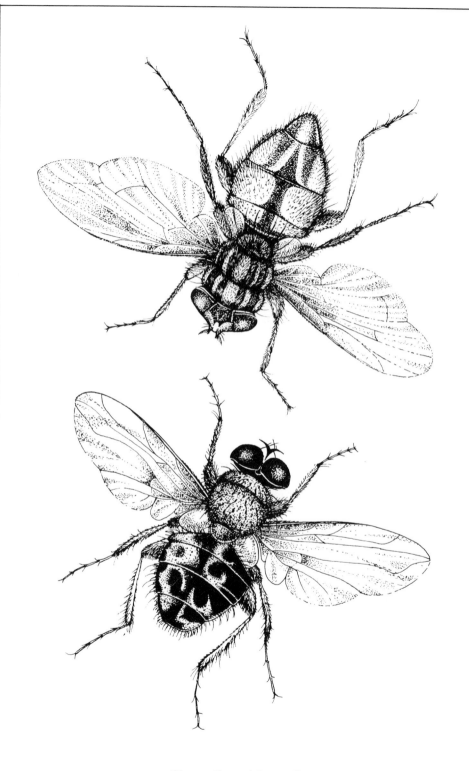

Cluster-fly and house-fly

One method of control was to keep these fermenting so fast that they were too hot for the flies: another was to put the horse, pig and human dung beneath the cow dung and thus out of reach of the house-flies, and a third method was to apply insecticide to the heap. In 1899 the farm was sold, being managed subsequently in a much more efficient manner: the manure heaps were more compactly made, breeding fewer flies, and the house being empty there was no human food in it so that few flies were attracted there.

The next anti-fly measure was the introduction of the 'Flit' type of spray between the wars in this century. These sprays started as white oil solutions of pyrethrum to which were added other insecticides, finally DDT and Allethrin being used, after the Second World War, to make a rapid and effective product. They were good, but the basic objection remained that they only killed the flies present in the house. The spray had much the same effect in reality as killing the flies one by one with a swatter, except that it was easier and quicker. New sprays with a bigger concentration of firstly DDT in them but now of more modern and less harmful insecticides, such as pirimphos and synergized pyrethrins leave a toxic residue of insecticide behind them on walls, ceilings and paintwork, which will kill flies alighting on them and so keep premises free for many weeks. But here again the flies have their remedy; in some places they have managed to produce races having a considerable resistance to the insecticide and thus make a brave attempt to keep up their numbers.

The most effective anti-fly measure was the virtual disappearance of the horse: this deprived the insects of their favourite breeding medium, and combined with a certain amount of house spraying and the treatment of refuse tips with BHC dust or malathion bait, has led almost to the extinction of the house-fly in towns, or has at least reduced the nuisance to very small proportions.

The farm at Bartons End has now a Grade A dairy herd and must keep down its flies. The manure is stacked on to concrete middens having larvae-traps all round them. They are shallow channels always full of liquid, which not only drown the larvae as they attempt to leave the stacks to pupate, but also allow the valuable liquid manure to be collected and used on the hops and soft fruit. Sprays are applied in the milking-shed and in the house. Flies are not extinct but their numbers are a tenth of what they were when first they were seen on the premises. No DDT resistance has yet been found and if it does arise other and newer insecticides will be used to overcome it.

Man has now proved to be the worst enemy of the fly, but this was not always so: at one time spiders and a fungus were, and these two still take a considerable toll. A spore of this fungus can germinate on a fly and the mycellium penetrate the body, killing it within a few days, when the insect becomes attached by fungus growth to whatever surface it last rested upon: fruiting bodies of the fungus then push out from the body segments, giving it a striped wasp-like appearance and produce a mass of spores ready to infect more flies.

'Where do flies go in the winter time?' was a song that rang round the world in 1922. Flies pass the winter mostly as pupae. As the weather gets colder their rate of development slows down and at the end of winter only the more favourably placed larvae and pupae will have survived, but their powers of reproduction are so prodigious that a few survivors are rapidly able to recolonize all suitable sites. In Chapter Two mention was made of the rapid build-up of elephant populations should all offspring survive, and the elephant is a slow breeder. This is a favourite exercise of the biologist wishing to impress his readers with the power of the life force and some impressive calculations have been made for the fast-breeding house-fly. A pair could increase to some 200 million billion in a season: this number would cover the entire earth to a depth of 47 feet with house-flies! It will readily be seen that the factor limiting the expansion of the fly population is more likely to be shortage of food than the activities of predators and, in fact, it is this denial of food to the larvae, by proper manure and refuse handling, that is the basis of the modern fly control system.

Cluster-flies occasionally cause alarm in houses; in fact, they once did so to Robert Dunchester until the matter was explained to him by his entomological friend from the research station. The larvae of cluster-flies are parasitic on earthworms: in the autumn many of the larvae kill their hosts and emerge as adults and seek a winter shelter. They frequently get into roofs where they cluster together and become torpid, looking very like house-flies, except that they bear a number of golden hairs on the thorax. Sometimes in the autumn they come into rooms where they are not able to survive long as the temperature is too high to induce torpidity, causing them to use up their food reserves in flying and walking about, so that they die before the winter is over. In the roof in spring, as the temperature rises, the flies become active and may get out of the roof into the rooms of the house, or they may even do this during a warm sunny spell in winter. One spring there were complaints of swarms of horrible flies trying to get into the bedrooms and all the windows and doors were shut to stop them. Actually they were trying to get *out* of the house and find earthworms on which to lay eggs. If Robert Dunchester had wanted to spare the earthworms of the neighbourhood he should have sprayed the flies with an insecticide, or if he just wanted to get rid of them then he had but to leave the windows open, perhaps turning up some fresh earth in the garden to tempt the flies out with a fine, wormy smell.

Blowflies of different sorts, the bluebottle, greenbottle and birdbottle came into the house from time to time, attracted mostly by the meat in the kitchen and larder where the two former are ever seeking to deposit their eggs. Meat exposed for a very short time can get blowfly eggs on it, even while lying in the dog's dinner-plate. The bell-shaped wire gauze meat covers are no final protection for meat, for even if they fit flat to the dish the female fly is capable of alighting on the gauze and dropping eggs through it on to the meat beneath, where they will soon hatch into maggots and consume it all. In fact, the voracity and numbers of fly larvae which can arise in meat led Linnaeus to state that a fly can eat an ox as fast as a lion.

The only other fly incident worthy of record is when some of the house-martin flies got into the bedroom of the Reverend R. Sheppard, who was having his portrait painted by the elder Robert Dunchester. They bit him and caused a certain amount of ill feeling as the bites were thought at first to be caused by fleas, which were not expected to be in a gentleman's house in those days: perhaps a successful artist was not really a gentleman after all. What had happened was that a house martin had died in a nest, upon which the swallow-flies immediately left the body and sought a new host; though they only like the blood of hirundine birds, they will experiment with anything they find, hence the attempt on the worthy cleric.

Mosquitoes and gnats are both summer and winter visitors. They belong to the same general family of insects as the flies. They feed on both man and birds and in the early days of the house were a frequent cause of fever in man, as one particular species of mosquito transmitted malaria fever—another cause of the high infant mortality of the age. The larval stages of mosquitoes and gnats are passed in water, feeding on particles of organic matter present there. By means of small brush-like appendages to the mouth-parts they collect together protozoa, bacteria, algae, fungal spores, pollen and similar material, work it into a ball and swallow it. As is well known it is certain of the anopheline mosquitoes that carry the malaria parasite. The common gnat, or *Culex pipiens* mosquito, carries no human parasites and in fact rarely bites man but is very annoying to birds, to whom it can transmit various diseases and filarial worms. Other kinds of mosquitoes and small gnats are trouble-some biters and disturbers of the general peace.

The larvae of *Anopheles* lie flat to the surface of the water, whereas *Culex* larvae hang head downwards from this layer. Both breathe through spiracles in contact with the air when the creatures are on the surface, having valves to close them when they dive—very like a submarine in fact. The anophelines swim tail first just under the surface in a series of random jerks, whereas the culicines wriggle about under the surface skin of the water. Both kinds of larvae plunge down deeply (if they can) when frightened, but they must eventually come to the surface again and eject their breathing tubes in order to get air. They breed mostly in the rain-water butts, but some come from puddles, ponds, water pockets in hollow trees and so forth. The pupal stage is in the water, and the pupa emerges by bursting the dorsal side with air pressure. The adult then rests on the old skin or on nearby plants, in order that the wings may fill and harden.

The dangerous anophelines and the harmless culicines are most readily distinguished by the position assumed by the adults when at rest, the former pointing its head down and tail up and the latter having its body parallel to the surface on which it is standing.

It is only the female mosquitoes that take blood from man and animals in Britain: the males live on juices sucked from fruit and flowers as the females also do at times. In the old days the house-gnats would haunt the dairy and suck the milk from the cream-setting pans; having to push the proboscis through a layer of cream to get at the milk beneath was to the gnat the same as pushing it through the skin of a vertebrate animal in order to get at the blood.

Apart from infecting one with malaria, mosquitoes can be very disturbing and annoying at night and there are many people who lie in bed in the dark listening to the whine of the creature come and go and wait for the abrupt cut-off, meaning that the insect has settled and may be about to strike—a silence reminiscent of the ominous cut-out of the flying bomb's roar. The musical may obtain some relief from this tension by knowing that the males (which will not feed on blood) have far higher-pitched notes than the females and that the note of the hungry female drops from the normal of F to that of D when she is fully fed.

Major Hackshaw started to take a greater interest in insects after he found the moth larvae attacking the corks of his port-wine bottles: he used to aver that his mosquitoes were particularly virile and voracious and that they enjoyed nothing more than a sup at his glass of port, provided that the glass was filled to the brim so that the creature could get at it. His wife Caroline maintained that this was done not so much in the interest of natural history, or to follow in the footsteps of the great Buffon, the famous eighteenth-century French naturalist who was popular in Britain, but as an excuse for the major and his friends always to have their glasses full.

It is unfortunate that in his diary the major only records the bare fact, which is quite possible, since port is only a sort of fruit juice, and not the subsequent behaviour of his toping insects; one would like to know if the clarity, pitch and steadiness of their flight notes were in any way affected.

The females of two kinds of mosquitoes and the common gnat used to winter in the house, mostly in the cellar, but they are also found in dark cupboards and the loft. They are not now so numerous as formerly because of the various insecticidal treatments given to the house, its drier condition, and the fact that the rain-water butts are treated to keep down the larvae. During the summer, whenever the Thompsons are spraying the roses against greenfly or giving any such insecticidal treatment in the garden, they always squirt a little of it into the water-butts as they go by, which is quite enough to keep down gnats and mosquito larvae there. A few drops of oil can be poured on the top of the water, which spread over the water surface, killing the larvae because it prevents them breathing. This film is soon lost when it rains if the barrels overflow from the top, as most of them do. Some of the wintering females may feed from time to time during the cold season; they convert this food to fat reserves, not to eggs, and come out in the spring to find suitable water sites on which to lay their eggs. Some mosquitoes do not lay in water but in depressions in the ground, hollow trees and so forth which will eventually fill with water and allow the eggs and larvae to develop.

The dangers of malarial infection in Britain are now almost extinct. To transmit the casual sporozoa the anopheline mosquito must bite an infected human, wait the necessary time for the parasite's life processes to develop within it and then bite another human. Though the casual mosquitoes are still common they are not so numerous as they were once; moreover, livestock are now found in larger numbers and the insects prefer to feed on them. Rooms in houses are now kept cleaner so that more wintering females are destroyed and finally malaria-carrying humans are now comparatively rare, as those returning home from the tropics infected are usually cured with such drugs as Paludrin and Mepacrin soon after arrival, so that we are fortunate in there being but little residue of the disease left in the country. This is another example of an organism (the malaria parasite) facing extinction because its numbers are falling below a certain critical level. When the house was first built 'the ague' was a common complaint and often a wretched labourer or servant, reprimanded and abused for being shiftless and idle, was merely suffering from malaria or the enlarged spleen that goes with it. Bites from mosquitoes (*Anopheles*, *Aedes* and similar genera) are often blamed on to the harmless house-gnats, but as the gnats are excessively irritating and harmful to birds these last live an easier life if the gnats are sprayed and killed; it is indeed no kindness to a cage-bird to leave it uncovered at night as it exposes the bird to the vicious attack of gnats.

There are, of course, a large number of small, biting midges to be found round the farm, but these rarely enter the house. There is also a parasitic mosquito which will not attack animals itself, but draws its meal of blood from other already gorged mosquitoes. The autumn and winter deaths among mosquitoes and gnats are very high, but their powers of reproduction are enormous and the vacant spaces for mosquitoes are quickly recolonized by the few survivors of the winter weather.

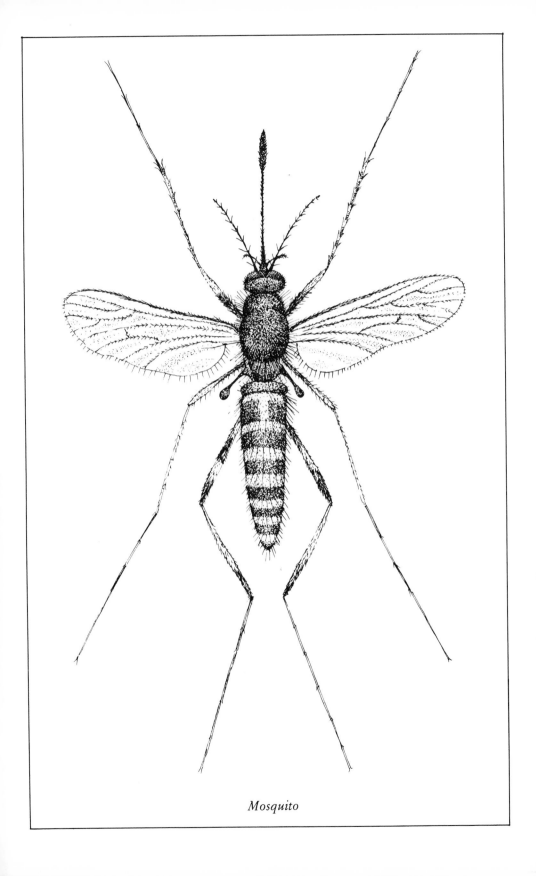

Mosquito

Ants, wasps, bees, greenfly, dormice, rats and earwigs were occasional summer visitors to the house.

The black or garden ant was the most usual invader of the kitchen and larder; at one time the tiny Pharaoh's ant was present and the small Argentine ant made an attempt to establish itself but without success. Two of these belong to the most advanced ant family, the *Formicinae*, leading the usual complex social life of these creatures with a caste system of specialists—workers, queens, drones and so forth, whilst the Pharaoh's ant is a myrmicine with an omnivorous diet. Maeterlinck has pointed out that the social development in the ant realm is not unlike that of the human world. The two most primitive families of ants (*Ponerinae* and *Dorylinae*) have but little caste distinction, are carnivorous and their life is thus comparable to the hunting stage of primitive man. The *Dorylinae* are not found in Britain but are the wandering ant hordes of the tropics, ceaselessly searching for animal food, who make no permanent homes, which again bears a resemblance to our own nomadic ancestors. Among the remaining three families (*Myrmicinae, Dolicoderinae* and *Formicinae*) may be found the parallels to the herdsman, farmer and city-dweller. Some ants carefully tend aphids (or greenfly), stroking them for the sugary secretion they give off, carrying them to new favourable sites to start a new colony and driving off the herd's enemies; such ants may be seen on almost any rose tree in summer.

The true farmer ants, which are not found in this country, cut leaves from plants, chew them up and then plant a special fungus on the resulting hot-bed, keeping it well weeded and trained so as to kill all alien fungi and to produce a good crop of the special 'bromatia' on which they feed. The omnivorous ants may be compared to the city-dweller, who will now eat almost anything. 'Go to the ant, thou sluggard,' says Solomon. 'Consider her ways and be wise.' Perhaps it is only from the point of view of the energy and determination they show that we should admire them, for their social organization has led them into as many undesirable activities, from the moral point of view, as has that of humans, to wit wars, slavery, child labour, parasitism, and over-eating.

The black or garden ant is found in good soil around houses. A new colony is started by a mated queen descending after her nuptial flight, breaking off her wings, digging into the soil and forming a cell in which she lays a few eggs. As these hatch the queen feeds the young grubs with excretion from her salivary glands, using up the substance of her food reserve and flying muscles in the process. The grubs grow, pupate and emerge as workers who wait on and feed the queen. As soon, as some workers appear the potential new colony's major difficulty is over, and provided they can find food, the queen can devote herself to the main task of laying eggs and killing any rival queens that may arise in the nest.

The workers, who are sterile females, forage for a wide range of foods, they kill small insects, take nectar from flowers, carry back small seeds and collect the sugary fluid exuded by their carefully tended greenfly. They send out scouts to look for food, and when an ant finds a good source she is able to indicate the fact to certain of her fellows, regurgitating a drop of the material in question and signalling with the antennae. These leader ants organize the work so that soon a long trail of the insects is winding to and from the new supply. It is these scouts that find their way to food in houses and soon the line forms to the pile of spilt sugar on the larder shelf or, if they need protein, to the cold Sunday joint. The trails, once established, persist for some time as the ants recognize them by the smell left on the track.

In the nest, frequently under the turf in the garden, the grubs are growing and pupating, carefully fed and tended by the workers. The white pupal cases are the familiar 'ant-eggs' which are all the time reaching maturity, splitting and adding to the population of worker ants. In late summer extra feeding produces a number of bigger pupal cases, containing the sexual forms—males and queens—which bear wings. As these hatch the workers retain them in the nest until the weather is just right—usually a fine day after rain—when the swarming phase takes place, from nearly all the nests at about the same time. Some establishments have a predominance of males and others of females. Mating takes place in the air and the disparity in the proportion of the sexes between nests results in a mixing of the blood lines. The males die soon after mating and the queens come to ground, discard their wings and attempt to form a new colony. During swarming ants get into the house, but though they sometimes frighten or disturb people they do no harm nor does the swarm last long.

Their method of forming new colonies is very different from that of the garden ant; although the winged sexual forms arise, they never fly and a new colony is made by a mass migration of several mated queens with a staff of workers, eggs, grubs and cocoons. The swarming phase is an immense loss to the ant community as very few of the females can hope to establish new colonies; it is an indication of the antiquity of the race of ants and the limitations of their system. Animals which are less prodigal of their adolescents, but take greater care of those available— such as bats and humans—are more successful.

The installation of central heating gave the tropical Pharaoh's ants an opportunity to install themselves in the house. They like warmth and can follow the pipe runs to hide their nests under floors or in the walls. They are not much seen, though sometimes they cause trouble to humans by getting into their beds and feeding on a wound, or taking eye or nose fluids for the food and water they contain. This makes them very objectionable in hospitals, a splendid environment for them because of the steady heating and supply of food there. The tiny, palish ants emerge and form their trails to and from food of all sorts, meat, cheese (protein), fats such as butter, lard and margarine, sugar, jam, honey and so on. The tracks will also lead to water sources, such as taps and drains, for finding water is one of their main problems. They are able to discover food in the most unlikely places; the piece of chocolate forgotten in a drawer will be found by these ants. Once they are established in a well-warmed building they are difficult to eliminate as they are so small and persistent, squeezing through the smallest cracks to establish colonies in quite inaccessible places.

It is no good spraying the ant trails to kill the insects found there as such small numbers will be eliminated that the operation is hardly worthwhile. The ants can be offered poison bait, but it must not be strongly poisoned or these wily creatures will soon leave it alone, either because those that take it die before they get the bait back to the nest, or because they discover it is not a good food for the grubs.

A lightly poisoned slow-acting bait is the ideal material against these house ants, because the workers will then take it back and feed it to the queen; once the queens are poisoned the colony fades away. The baits are put into waxed pill boxes or little tins with holes in them and have frequently to be renewed; they are usually poisoned with sodium fluoride.

Of course the fundamental method of ant control is to find and destroy the nests. Those of the Pharaoh's ant are so hidden and so inaccessible that it is almost impossible to find all of them; it means tearing down walls and taking up floors to find even some of them. It is, however, easy to find the nests of the black ant, simply by following back the long trail of workers, which must eventually lead there. If the nuisance warrants it, for otherwise it seems a pity to destroy such diligence, nests can be eliminated by lifting the turf, and sprinkling in a moderately volatile insecticide powder, such as pirimphos.

The system cannot be used with the other ants; should an attack prove serious, recourse is made to baits placed near their runs and to insecticidal varnish painted in six-inch bands where the ants are likely to cross.

Another visitor was the earwig, a plague of which was found in the house the summer after it was built and arose from the disturbance to the local population on the site. As plants such as wisteria and a vine were trained up the sunny walls, so was access made easier for earwigs and to this day they are frequent summer and autumn visitors. They do not breed indoors, but in soil under stones, posts, in wall footings and in the damper parts of the creeper on the walls. The female insects are remarkable in the care they take of the eggs and the newly hatched young, fussing over them somewhat in the same way as a hen will over her chicks. It is remarkable that the eggs become mouldy and do not hatch if the female neglects them.

Earwigs feed on both animal and vegetable food and can be particularly objectionable to the gardener, especially if he grows chrysanthemums, dahlias and peaches, as they can destroy many choice blooms and ripe fruits. They dislike the light and so feed at night: they have one generation a year and are mature from August. The large wings are seldom seen; they are very delicate and have to be carefully folded up with the help of the familiar tweezers on the insect's final segment.

Nearly all the visitors to the house were insects, but there were occasional other arrivals, for the edible dormouse was seen on a few occasions and even gave rise to some alarm as it is a most noisy creature, blundering around in the dark looking for food, knocking things over and generally creating a fuss. Many of the breakages, noises and other happenings attributed to the little people or poltergeists were really the results of the occasional visits of the fat or edible dormice.

Human visitors constantly came and went and had much effect on the life of the place because social standards required certain patterns of behaviour, cleanliness, furnishings, special food and so forth, which all had an influence on the life there.

Visitors who were found, like the humans, in both summer and winter were the wasps, for during the summer the premises, particularly the kitchen, would be invaded by the worker wasps, while during the winter, first in the thatch and then under the tiles, numerous queen wasps would be found in their winter hibernation. There are seven kinds of wasps in Britain. The hornet and Austrian wasps are comparatively rare. The common, German and red wasps build underground nests, whilst the tree and Norwegian wasps hang their homes among the branches of trees and bushes; the hornet uses a hollow tree as a nest site.

Queen wasps feed heavily before their nuptial flight and after mating seek a dry, safe, reasonably warm spot for hibernation, which is why they so often come into houses, frequently being found clinging to curtains and under roofing tiles. With the warmth of spring the queens emerge to seek nesting sites. The subterranean nests are the more common in Kent; a queen, having found a suitable hole in a bank, often an abandoned mouse burrow, proceeds to make a cell at the end of it, and line it with 'wasp-paper'. This she prepares by rasping off weather-worn but sound wood and chewing it up with saliva. It is first spread under the roof of the cavity and from the centre of this disc, or inverted saucer, a stalk is dropped which opens out to form another disc, or lower storey. On the lower side of this the queen makes the cells, hexagonal like a honey-comb (in fact bees are really only wasps that have lost the carnivorous habit) and open at the bottom end. In these she lays and cements down the eggs, some thirty in number, feeding the young larvae until they are ready to pupate, which they do hanging in the cell and closing the bottom with silk.

Up to this point the queen has been doing all the work herself—and an immense task it is to find and prepare the site and the paper, then feed herself and the grubs. All the grubs emerge as workers, that is immature females, who immediately set about extending the nest and caring for the young, the queen subsequently devoting herself solely to egg-laying, fertilizing them from the stored sperm she obtained when mated on the nuptial flight. Successive lower storeys are added to the nest as its strength of workers grows and the whole is protected by the formation of a paper envelope around it. The cells are used again at least once, and some of the older ones carry a third brood as well, so that though a nest may contain some 11,000 cells it may well have produced 25,000 wasps during the season and have a strength of some five or six thousand at the end of it. Towards the end of summer bigger cells are constructed and the grubs in them are better fed, becoming the mature females, or queens, of the next generation. At the same time numerous males are produced from unfertilized eggs, and some mate with the queens on the nuptial flight, the cycle then being repeated.

The food of wasps is both animal and vegetable. They take flies, butterflies and other insects, and meat and fish from the kitchen, which they partially masticate before feeding it to the young. As is only too well known, they also like sugary fluids, ripe fruit, honey dew from aphids and the nectar of flowers, but they are not as successful as bees in obtaining this last as their mouth-parts are not long enough.

Wasps were attracted to the house for the same reason as so many other animals were; man made it much easier for them to get their food; rich proteins and splendid carbohydrates would be left lying around in large quantities for the taking, in addition to which large numbers of insects would also be drawn to the premises by man's activities, and also be of use to the wasps. No wonder that the presence of man in a neighbourhood always leads to an increase of the social wasps. Had they only learned not to sting their benefactor they would have been yet more successful, for though killing workers one by one on window panes makes very little difference to their numbers, the unpleasantness of the stings leads man to seek out and destroy the nests. At first he would burn sulphur in the mouth of the nest; today he puts in a spoonful of insecticide powder which is carried into the cavities by the successive entries of the workers.

William Daker spent a lot of time studying wasps, which would have been better applied in looking after his farm. He noted one thing which enabled him to astonish his friends. Male wasps or drones do not sting and can be distinguished from workers by having seven instead of six abdominal segments and longer antennae. He also found that only the males seemed attracted to the autumn flowers of the ivy, a plant which

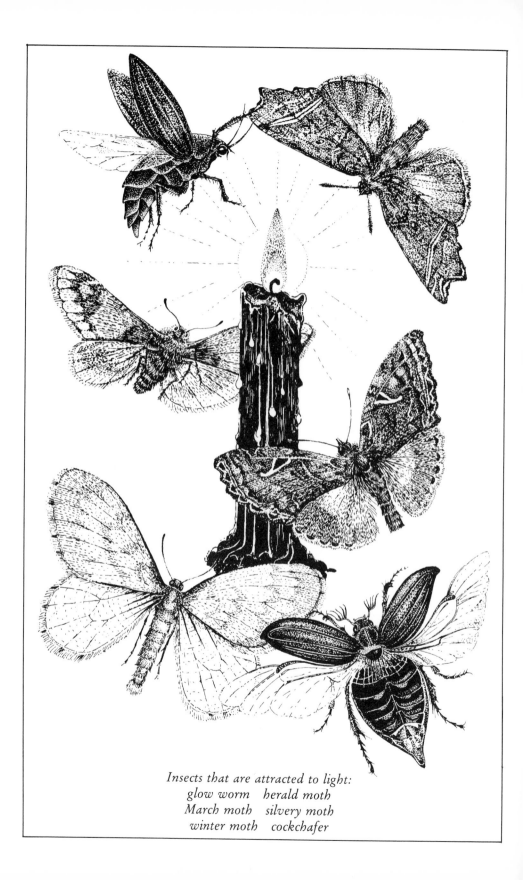

Insects that are attracted to light:
glow worm   herald moth
March moth   silvery moth
winter moth   cockchafer

was becoming far too common as the place got more and more neglected.

'I can charm these wasps,' he would say, running his fingers among them as they clustered over the ivy heads or picking a sprig complete with the insects and putting it in his buttonhole, actions which never failed to give rise to cries of surprise and astonishment. However, he got a little too confident because on one occasion at least there were also some workers present, who at once stung him, an event he attempted to conceal with the fortitude of a Spartan. Another of his inventions was to tie a piece of white cotton to a wasp in order to slow down its flight, and thus get an idea of the whereabouts of the nest.

Other summer visitors were those insects attracted to lights in the house. These were mostly moths and the assassin bug which I have already mentioned. The most frequent moths found round the lamps are the herald, silver-Y, winter and March moths. Cockchafers and male glow-worms were also drawn and, in fact, a whole host of other insects as well, too big to enumerate at length.

The reactions of insects to light can be very varied; some, for example fly larvae, will avoid it continuously: others seek it at one stage of their history and avoid it at another, as for instance the adult *Vanessa* butterflies and their caterpillars. The insects drawn to a lighted window were far more numerous when the night was dark than when it was lit with bright moonlight, because a light-attracted insect is drawn by a point of light and not by a sheet of light, which is the reason some moths come to the candle and not to the sun, for the sun's rays are parallel, whereas those of the candle spread from a point. Insects which are drawn to lights are so affected because in the wild they use a distant source of natural light—the sun or the moon—to orient themselves, for, by moving so as to keep the angle of light impinging on the retinas of the compound eyes the same, they move forward in a straight line, at any rate over a short period of time. One remembers as a child walking along a straight road that one could catch up with a lamp-post but never with the moon.

If a source of light is near, the insect will not know this and will proceed as if it were a distant, natural light; it will keep the light angle on its retinas the same, which means it must be constantly turning towards the source, resulting in a spiral approach to the light until it hits it and

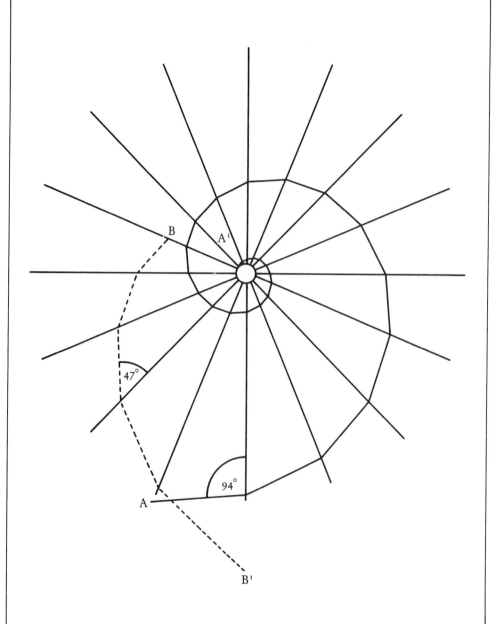

FIGURE III (i) *Insect orienting on a light*

An insect does not realize a light source is near and in orienting itself on
one acts as if it were distant: it consequently keeps the angle the light makes
on its compound eyes the same, spiralling inevitably towards the source
(A—A¹) or away from it (B—B¹), according to its starting angle.

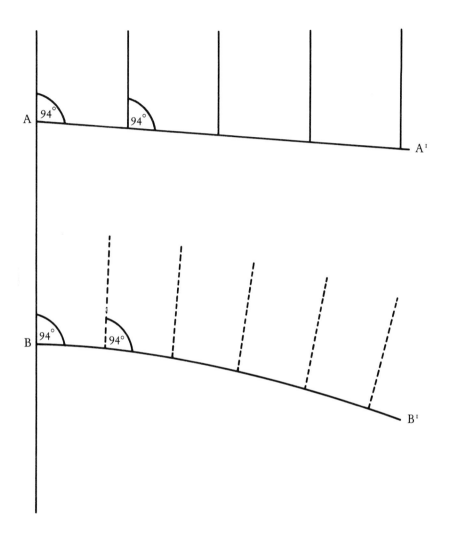

FIGURE III (ii)   *Orienting on a distant source of light such as the sun or*
moon

The rays of a distant source of light are parallel and an insect starting at A
and keeping a constant angle with this source of light, in the above case 94°,
will move from A to A¹, that is over a short period of time.

   In practice the sun appears to move 15° in an hour, which is represented
by the successive dotted lines from left to right above, which are at an angle
of 3° to each other, and represent intervals of 12½ minutes of time. This
means the insect moves course every 12½ minutes and would proceed from B
to B¹ keeping the same angle, 94°, all the time. In practice, journeys of an
hour's duration in what the insect thinks is a straight line are of course
very rare.

usually is killed. The line traced by the insect is actually a logarithmic function of the path the creature was pursuing when first it started to orient on the light. If you think of the light as emitting a number of rays like the points of a mariner's compass, the insect as pursuing an original course of other than a chord between two of these rays and that the creature keeps this angle constant every time it crosses a ray, then it must spiral towards the centre if this angle is greater than the chord-angle and away from it if it is less (see Figure 111). Some insects are repelled by lights, perhaps for this reason, but many moths will spiral towards a light, as we can easily see in summer when they flutter around a lamp.

Winter visitors to the house included some of these *Vanessa* butterflies, the most numerous of the genus being the small tortoiseshell, though the peacock was frequently found as well. The caterpillars of these insects feed on nettles and so are attracted to man's houses, for he tends to scatter nitrogen around his dwellings, a substance much appreciated by those plants. The small tortoiseshell has two generations a year and the peacock only one. In the autumn the adult butterflies seek places in which to hibernate; old rabbit burrows, a hollow tree, a pile of brushwood, may be used, but favourite places are houses, particularly the lofts, where in the spring the tortoiseshells, peacocks and sometimes red admirals may be seen fluttering about in an attempt to get out, find some nettles and start the cycle again. In the late summer and autumn these decorative creatures are much attracted to buddleia trees and michaelmas daisies in gardens, stocking up on the nectar still to be obtained from these plants to form the big food reserve they must accumulate to pass the winter and still have considerable substance left in the spring to lay the eggs for the next generation. In the case of the other *Vanessas*—the admirals, painted lady, and the large tortoiseshell—it seems unlikely that the wintering adults in this country survive in sufficient quantity, or have enough energy left to lay eggs and that the population is in effect carried on by migrants coming in from France, on the south-west breeze. The small tortoiseshell and the peacock may also come in this way, as does another common butterfly—a great nuisance to man—the cabbage white.

William Daker found it fortunate that so many of these pretty *Vanessa* butterflies fed on nettles, as he could always offer this as an excuse for letting the plants appear in such profusion around the house.

Another winter visitor in considerable numbers was the ladybird, whose very name suggests that its good qualities towards man, in eating many kinds of insects, have been known for a long time. In both French ('bête à bon Dieu') and in German ('Märienkafer') it is similarly recognized. These insects lay batches of yellow eggs on a number of plants, which soon hatch into hungry, grey-black 'niggers', feeding voraciously on plant-lice or greenfly they find living on the plant. The 'niggers' pupate in due course and emerge as adult ladybirds, who are equally hungry for greenfly. In the autumn the adults frequently move into houses and flock together, where they sometimes cause alarm. They are, however, doing no harm and should be encouraged to depart when the warm weather arrives, though it is unnecessarily cruel to speed them on their way by suggestion that disaster has overtaken their homes and families.

Visitors to the cellars have already been mentioned a few times, together with the animals found there—the firebrats, silverfish, cockroaches, house moths, spiders, beetles and so forth. There were two different cellar communities; when it was first built it mostly housed beer and wine barrels and was very damp. This attracted aquatic insects, mosquitoes, slugs, snails and toads. Later on the central heating boiler was installed and dried up the place so that a quite different fauna became established, those liking warmth and able to stand some degree of dryness in the surroundings.

In the first community both the common toad and the natterjack were found from time to time, the latter much more frequently as it ventures, usually in pairs, further from the water than the former. The eggs of toads are laid in water, in much the same way as those of frogs, and the creatures pass through the same tadpole stage; the eggs, however, are laid in long strings rather than in gelatinous masses. The adults feed on spiders, insects, slugs, worms, beetles, cockroaches and other night-moving creatures, catching them as they move by darting out their long tongues, and on this diet they live to a great age. The natterjack is much smaller than the toad and can be distinguished as well by having a pale line drawn down its back; they haunt much drier habitats than the common toad though they must have some water and, like the toad, return to water in the spring to mate and breed. The natterjack is now comparatively rare. Both toads and frogs have a great sense of direction and locality; they find their way to water by smelling it, the males moving off first and then calling the females to it. After breeding they frequently come back to their original locality.

Here I must leave the animals in the house with just a brief mention of a few more. They were nearly all visitors. Greenfly grew on the pot plants and often formed the first meal of the ladybirds moving out of hibernation. Soft scale was also found on the plants and tarsonemid mites attacked the crowns of cyclamens. Lacewing flies sometimes attacked these greenfly, the delicate creatures laying a number of white eggs on thin stalks among the leaves; the larvae on hatching set to work to eat the aphids. Earthworms were sometimes present in these pots. Blossom beetles came into the house with cut flowers and a whole zoological garden of animals—woodlice, millepedes, centipedes, slugs, snails, sping-tails, beetles and so forth—came in with the firewood.

Doctors occasionally brought in leeches to bleed their patients but the most remarkable creature to visit the house was one, apparently abundant, of a metaphysical nature. When the elder Robert Dunchester was rising towards prosperity, he acquired a valuable Welsh house-keeper who kept the place as bright as a new pin and charmed his wealthy sitters. The war against insects preyed on the poor woman's mind until she nearly became distracted by the attacks of small invisible insects 'living under the skin'. Needless to say they were entirely imaginary—such occurrences are quite common still—and fortunately Myfanwy was able to convince her 'treasure' that painting her all over with a solution of methylene blue, supplied by the local doctor, so that she looked like one of her savage ancestors, was an effective and permanent cure, though they had some anxious moments with her during insect incidents with the Belgian refugees in the First World War.

Chapter 13

Past, Present and Future

What is a man,
If his chief good and market of his time
Be but to sleep and feed? A beast, no more.
Sure, he that made us with such large discourse,
Looking before and after, gave us not
That capability and god-like reason
To fust in us unus'd.

WHEN BARTONS End was built a revolutionary disturbance of the life in the neighbourhood took place. For example the spider community, living on the insects and mites on the site, was destroyed as were also those animals living on the trees used to make the house, and in the forest environment that they formed. Clay was dug for the bricks and lime burnt for the mortar, both actions involving disturbance to the surface life of claypits and limestone quarries. In exchange the house rapidly filled with a life of its own, which was possibly as great as the life destroyed in building it, but was very different in nature.

We have followed the development of some of these creatures and it is possible roughly to measure these changes, for, as Lord Kelvin said, one can only understand a thing when one can measure it. There are, however, a number of different ways of measuring life—for instance, we can count the numbers of individuals of the different species existing at different points in the history of the house, or we could calculate their weight, or the weight of the vital animal protein they contain. The rate at which they live—their oxygen consumption, or metabolism, can be measured and some consideration could be given also to the numbers of generations they have gone through in the course of the house's history. Finally we can attempt to assess the quality of the life in the house from the points of view of dominance and moral values, or the relative happiness of the animals there. Each of these methods will show up some differences and similarities between all these diverse creatures.

Their relative numbers and weights are best shown in a table, which has been done on page 227 for the principal animals, at fifty-year intervals. Some creatures such as spring-tails and mites are too small and too numerous to count and are shown in the table by a series of + signs, a greater number of signs indicating a bigger number of animals. The figures are also set out in graphic form for humans, and wood-borers in Figure 1 and for spiders in Figure 11, pages 55 and 103. Even without those innumerable small creatures the insects and spiders were obviously the biggest group in the house, passing from 1,092 and 3 + in 1558 to a peak of 3,827 and 7 + in 1860, and a present level of 452 and 2 +, whilst the vertebrates ran from 22 at first to the peak of 160 in 1910 and a present level of 20, no great change from the starting point, though this figure does exclude the toads.

We hardly need such a table to know that the humans dominated: though not the least numerous creature, as one might expect them to be, they were nearly always the heaviest accumulation of life, except, of course, when the house was empty of them. Their weight varied from 96 per cent in 1860, when Emmanuel Burrows was tenant and had a large family, to the present 72 per cent, again excepting 1910 when the total weight of animal life in the house was only $6\frac{1}{2}$ pounds. The protein content was roughly the same percentage of the animals' weight in all species, so the humans dominated in this field as well.

The rate at which they used oxygen is an interesting facet of the life because this varies enormously between different animals, when we consider it on an equal weight basis. Animals obtain their energy by oxidizing food within their bodies, virtually burning it there, and warm-blooded ones also maintain their temperature by doing this. As he

TABLE I
Approximate numbers and weights of principal animals living at Bartons End in certain years
Census day 1st June

Date	1558	1610	1660	1710	1760	1810	1860	1910	1958	1970	1980
Humans	7	13	6	7	8	4	21	—	2	18	21
Dogs	2	2	1	6	2	1	2	—	4	1	3
Cats	2	1	1	3	2	5	2	—	2	1	1
Mice	5	10	12	20	0	2	20	120	2	—	—
Bats	—	—	1	1	—	—	20	20	—	—	—
Birds	6	12	25	10	8	15	50	40	12	12	14
Spiders	500	600	1,600	600	600	1,200	1,500	1,400	200	100	50
Wood-boring beetles	50	150	300	700	600	500	700	1,200	20	50	10
Fleas, lice*	200	500	400	300	200	150	250	100	80	50	10
Bed-bugs	—	—	1	1	—	—	20	—	—	—	—
Clothes moths	20	100	200	80	60	80	20	100	20	20	15
Cockroaches	—	—	—	—	—	—	150	—	10	20	—
Flies	300	450	600	750	600	700	900	50	100	100	100
Spring-tails ⎫ Silverfish ⎬	+	++	++	++	++	++	+++	+++	+	+	+
Mites	++	++++	+++	++	++	++	++++	+++	+	+	+
	1,092	1,838	3,146	2,376	2,090	2,657	3,827	3,010	452	302	214
	+3+	+5+	+7+	+4+	+5+	+4+	+7+	+6+	+2+	+2+	+2+

* Of humans, dogs, cats, mice and birds

Weight in lbs.

Date	1558	1610	1660	1710	1760	1810	1860	1910	1958	1970	1980
Humans	739	1,184	709	851	736	472	1,683	—	251	1,640	1,913
Dogs	46	62	30	45	56	30	42	—	84	10	32
Cats	24	12	11	34	22	55	24	—	14	8	8
Mice	0.36	0.72	0.86	1.43	0.72	0.14	1.43	—	0.14	—	—
Bats	—	—	0.01	0.01	—	—	0.42	4.74	—	—	—
Birds	0.24	0.48	0.99	0.39	0.40	0.59	0.42	1.59	0.61	0.5	—
Spiders	0.02	0.03	0.08	0.02	0.03	0.06	0.08	0.07	0.01	⎫ negligible	0.6
Insects and mites	0.05	0.10	0.13	0.16	0.13	0.13	0.21	0.13	0.03	⎬	—
	809.67	1,259.33	752.07	932.00	815.28	557.92	1,751.62	6.53	345.79	1,658.5	1,943.6
Per cent human	91%	94%	94%	91%	90%	85%	96%	—	71%	99%	98%

*Of humans, dogs, cats, mice and birds

breathes in and out man draws cold air in and exhales warm air and about one-fifth of the heat he loses is due to this process, most of the rest being radiated away from the skin. In a well-covered dog, nearly all the heat lost is through the mouth, which is why a dog pants so in hot weather: it is his only method of getting rid of heat and keeping cool. The heat and energy requirements of an animal are measured in calories, a calorie being the amount of heat needed to raise a litre of water from o to 1° Centigrade. A man asleep uses them at the rate of 1,500 a day; doing heavy work he needs 3,500 and sedentary work uses 2,500. If we take a figure of 3,000 calories a day for a 70-kilo man (11 stone 3 lbs) he is using 41 calories a day per kilo of body weight, this providing for all his movements and the keeping up of the body temperature. The calories are obtained from his food, a gram of carbohydrate or of protein giving him 4.1 calories and a gram of fat 9.1. One might imagine that a cold-blooded animal, such as an insect, would need far fewer calories to keep it active, but strangely enough the reverse is the case. It is not difficult to measure the oxygen consumption of insects and from this to calculate their calorie needs. Naturally there are very big differrences according to the insect's state and its activities. The table on page 229 gives some of these figures.

A bee at rest uses about fifty times more energy for its weight than a man and when flying some two hundred and fifty times more. An active bat uses fourteen times more, as we have seen in Chapter Six. The sparrow, mouse and bat are much the same size and have similar energy requirements. In general the larger the animal is, the less food per kilo of its weight does it need; for instance the horse, much bigger than a man, only needs about two-thirds of the food per kilo that a man does, whilst flies to the equivalent weight of a man would need about nine times his ration. It may well be that man's comparatively economical use of oxygen is one of the reasons for his success in life. On the other hand, man has been more successful than the horse, which uses its food to better advantage. With all these animals if one works out the calorie intake per unit area (per square metre for example) of body surface it will be found that the figures are very similar for all of them, as one might expect.

The numbers of generations of the different animals in the house also showed wide variations. There were in effect twenty-one generations of

TABLE II

Metabolism of different animals at Bartons End

Animal	Air temp during estimation °C	Weight gms.	Calories per kilo per day	Notes
Bee, in flight		0.09	10,500	W
at rest			218	W
Fly, adult	20	0.021	357	K
,,	30		705	W
larva	20		150	K
,,	30		240	W
pupa	20		30	K
,,	30		58	W
House gnat (*Culex*)		0.008	65	E
Cockchafer	20	1	82	R and R
	20		108	B and S
Cockroach	20	0.7	31	V from CO_2
Mouse	7	13	639	T
Rat	16	117	227	T
Bat, active			700	V-F
torpid			58	V-F
Sparrow		22	755	T
Dog		18,200	46	L
Cat		2,500	80	L
Horse		450,000	27	L
Man,		70,000	41	
basic			24	

w Wigglesworth: k Krogh: e Ellinger: r and r Regnault & Reiset: b and s Batelli and Stern: v Vernon: l Loewy: v-f Vesey-Fitzgerald: t Tigstedt.

5.047 calories per litre of oxygen.

humans, for this is the number of different owners or tenants, though they were not all lineal descendants from previous occupiers. At the other end of the scale there were some 4,200 generations of some of the mites. Intermediate numbers are seen in the 150 generations of furniture beetles and the fifty-five generations of dogs, where again in this last case they were not by any means direct descendants, but represented a number of different varieties, purchased and introduced one way and the other at different times. If ownership of the house may be said to belong to the direct descendants of original occupiers then the furniture beetles have the best claim to it, for they were in the original timbers of the house and were the first occupiers. A few of the direct line are still there after the many generations that have passed after the place was built. If we take a human generation at twenty years, the furniture beetles' descent is equivalent to 3,000 years of human succession, which would date Bartons End back to 1116 BC—a period when the site was covered with forest and the life was very different; when Stonehenge was just a thought or still young and men kept to the highlands and plains rather than the dangerous, mysterious and sinister woods.

We have taken 4,200 generations for the mites: a similar number of generations for man would put us back in the second ice-age, well before modern man, about 84,000 BC. This was the time of the Acheulian man, the being who had started to improve the almost natural flint tools (the eoliths) of his forebears, by making the more massive shaped, unshafted hand axes, sometimes called 'fist-flints'.

Evolutionary forces were at work at Bartons End as elsewhere and 4,200 generations, or very considerably less, are quite enough to produce well-marked effects, as may be seen in man himself, for the Acheulian man was an ape-like creature very different from ourselves. It is difficult to assess the evolutionary changes that took place around Bartons End in its four hundred plus years of existence, as some of these may be truly genetical and some a reversible response to changed conditions. Man became taller, healthier, longer-lived and more and more dependent on his increasingly complicated artifacts. The house sparrows became more closely attached to man, frequenting him more in the town than in the country. They may well have to desert him or decline, as man has almost given up the horse. Wood-borers were continually experimenting with new imported timbers: spiders have adapted themselves to the house and are now seeing if they can live in the new, warm, dry conditions. Most of the dry-litter community have developed an affinity for man's goods so great that today they can scarce do without them. Only the cats, except those deliberately bred by man, have scarcely changed at all, retaining the attributes of the animal first brought into the house by John Barton.

The food of all these animals was, and is, basically the same, but the area from which this food was drawn showed great diversity. At first the house martins drew nourishment from the most scattered area, for they travelled twice a year right across the world whereas the Bartons fed entirely from their own farm. The wood-borers were perforce confined to the house as larvae, though some adults ventured afield to feed on flower pollen. The spiders fed only in the house, for the young which migrated were unlikely to return. Lice and internal parasites ventured abroad with their hosts and some of the fleas did so, though the majority stayed in the bedclothes or other nesting material. As time went on, the actual area from which the humans drew their food at first decreased, as farming became more efficient, and then increased again as more and more meat was eaten—plants consumed as cereals, beans and so forth give more food per acre than when converted into meat and then eaten. As the weight of the crops harvested per acre increased so in general did the farmers become more prosperous, though at times a big crop, particularly in the case of hops, meant less money for the farmer as the price was so much lower per ton of produce reaped. From the last half of the nineteenth century the districts from which the food was drawn were much extended. The grain for bread was imported first from France, then from North America and Australia and, with refrigeration, meat followed from much the same quarters until today people, with an occasional jar of caviar, draw their supplies from the globe itself, including Russia and her colonies, and thus far outstrip the extensive zone which the house martins are still using.

The approximate daily calorie intake of the house at three different periods is given in the table below, using the same calorie rates as given in the table on page 229 and averaging the spiders and mites at 1,000 calories per kilo per day.

The calorie requirements of the house have halved because there is much less weight of life in it; the protein requirements have not dropped in the same proportion for two reasons: more animal protein is now eaten and nearly a quarter of the present calorie need is for 'Chubbs' (the Chow bitch whose real name is 'Lady Teablossom of Bartonsend') and her puppies—much of which is being supplied by high-protein meat.

Some of the six per cent of the weight of the life was composed of nitrogen combined with carbon and hydrogen to form this protein. The nitrogen cycle is the key to the life of the house and to the improvement

TABLE III

Total calories used per day (1st June)

	1558	1760	1958
Man	13,735	13,694	4,674
Dogs	966	1,150	1,748
Cats	880	800	480
Mice and bats	104	208	41
Birds	82	136	208
Insects, mites and spiders	34	209	14
Total	15,801	16,197	7,165

of the living conditions of the animals. Let me recapitulate the story: nitrogen is absorbed from the soil as nitrate and is formed into protein in the sunlit, green leaf. The soluble nitrogen, under natural conditions, reaches the soil in three ways: from the decay of vegetable or animal matter, from thunderstorms combining the nitrogen of the air, or from the action of certain bacteria found in the root nodules of leguminous plants. The farmers at Bartons End started to get bigger crops when they managed their land well, which was in effect either when they added more nitrogen, phosphate and potash to the land, or when they conserved what they already had, or used it with greater skill. Until about 1835, in common with the rest of the world as a whole, they could not increase their total store of nitrogen in the soil, for that obtained from thunderstorms and clovers only made up their losses; all they could do was move it around, transferring the fertility of one piece of ground to another piece as compost and farmyard manure were moved.

Chile nitrate, about 1835, and synthetic nitrates (obtained by taming the thunderstorm in a factory) greatly increased the district's prosperity at the turn of the present century and enabled Alfred Hackshaw to sell the farm in 1899 without the house, a thing it would have been almost impossible to do at any earlier stage.

To a considerable extent Bartons End is a model of the world outside. Taking life as a whole over the course of evolution, the only characteristic of the progressive changes that is common to it all is its constant tendency to expand — its effect of filling up all vacant spaces, even those formed by the very expansion of life itself. Life expanded when plants started colonizing the seashore and then the land, after which vertebrate amphibians started to fill up this new expansion by climbing out of the sea and eating the plants. In the same way this house was an expansion of opportunity for life: built for man, it filled with many other creatures who found they could use the vacant spaces not used by man, or not sufficiently well protected by him. The number of individuals now living there and their weight is less than when it was built: the dominant creature is less annoyed by parasites and there is now a vacant space in the house, which was at one time filled with life. This may be only temporary, for it may soon fill again, probably with human life, for as the population of the country increases and as the individual incomes of this population tend to move nearer to the average through such measures as taxation, higher wages and subsidies, so fewer and fewer individuals can afford to live in a large house. As the Dunchesters had no children and lived from a small investment income and the pen they thought they might soon have to let rooms, or divide the house into two and let one part, which would have again raised its content of life to something like its original level. However, in the event, Robert Dunchester died in 1958, aged forty-two years. Louise, his wife, inherited an interest for her life, but she remarried and Bartons End reverted to the three Thompson children. They overcame the problem of the cost of upkeep by converting it to an old people's home. Thus the house remained intact. But it was not a house in the strict sense of the word. Nobody slept there; it was a gathering place during the day for the patients.

We have now discussed the more tangible aspects of the matter—the numbers of individuals, their weight, descent and food requirements. Is there any way of measuring the quality of the life there? It is a difficult if not impossible thing to do. There are certain standards of human conduct or behaviour which enable us to say that on the whole such and such a person is good, bad, happy or unhappy, but these are subjective valuations dictated by personal opinions of the observer, which will vary from person to person, not objective, verifiable records such as the number of beings, their weight, food requirements and so forth. This is the constant dilemma of the philosopher and economist—the measurement of the quality of life or the 'utility' of goods or actions. In so far as a human assessment is possible one could attempt to weigh up the qualities of the human life by getting sufficient people to consider the facts and present an opinion on them—a Gallup poll in fact—but the exercise is hardly worth while, for the answer would be but human and the opinions of the pollsters would cover such a vast range—from 'Oh King, live for ever!' to Solon's 'Call no man happy till he be dead'—that any average arrived at would have no meaning, even were the opinions objective. Of course the economist is more fortunate in measuring utility of goods and services to man, because he can to some extent assess them by noting how much time, money or energy is spent on these different activities.

We cannot measure the quality of human life for the majority of us have time to spare over and above that needed for the obtaining of the necessities of life, which we employ in a vast range of activities for the satisfaction of our souls' inner needs.

The thought processes of animals are much less complex than those of man—the need for them being replaced in most cases by the existence of instincts. We could attempt to measure the quality of the animal life in Bartons End, finding possibly that the furniture beetle obtained more satisfaction from its life than did the spring-tail from its precarious existence and so forth, but to do so we have to set a human standard and postulate that the various animals react to circumstances and obtain satisfaction if not in the same way as we do, at least in similar ways, which is not the case. We must admit then that we cannot measure the quality of life in the house with any objectivity; we can only make personal subjective judgements of it and admire the earwig for her mother-love, the painted lady for her beauty, the dog for its devotion, and the house martin for its flight, thus committing the greatest crime in the modern biologist's calendar—anthropomorphism—or else assess the quality of their life in terms of its bearing on the future of the species—the adaptation of the animal towards its survival as a species, a subject already adequately covered in other books.

The existence of these instincts in animals may well have a bearing on contentment, because the more a being is controlled by instinct the more it will always know what to do. It may not do the best thing for itself under the circumstances but it will be in no doubt as to what should be done. The less instinctively controlled animal, on the other hand, may frequently not know what to do, as witness the admittedly somewhat extreme case of Aesop's donkey who, unable to reach a decision, went hungry when placed, equidistant, between two equally delicious bundles of hay. The higher animals have to think things out, and it is not finding answers to certain problems that leads to unhappiness, particularly in man.

We have thus another dilemma—is it best to react to a problem instinctively and be happy, but probably wrong (and one can pay for an error with one's life), or to worry over the problem and be unhappy if one does not solve it? This dilemma may have led to Solon's remark on happiness.

Human emotions and feelings are undoubtedly descended from those of animals; the human has gone so far ahead because of his one great, early discovery—the passage of time: the appreciation that there is a past, a present and a future. Animals live mostly for the present: their appreciation of other time states is very little, for, though the house martin and the wasp may learn a route, they indulge in few modifications of behaviour because of some past experience, for such changes they do make because of these are relatively simple and require but little thought. Birds build nests as a provision for future young; mice store food against winter scarcity but this is an instinctive pattern, not a thought-out activity. Man made his first step on the road to success when he used his leisure (in contrast to the animal who just played or slept) to think of the past, the present and to devise improvements for the future, developing not only tangible assets from the flint axe onwards, but a number of very difficult intellectual concepts as well— language, the zero, writing, money, to give some examples—all things which were difficult to invent and are of immense value, but which like so many things can be used by anyone once they have been invented. The artifacts helped him to live; the concepts enabled him to pass the information on, both to his fellows and to unborn generations and to become aware of yet more, wider and higher concepts—the arts, morality, science, ethics and religion.

The moral qualities have a biological value for us and though they are rare in the animal world yet they are not unknown. Mother-love for instance has immense survival value among a large number of animals. In the case of insects (the earwig and some hymenoptera being notable exceptions) the female lays many eggs in a suitable spot near food and then abandons them: usually enough survive to continue the race. Other animals find fewer offspring with more parental care a better way of ensuring their posterity; hence if the parents, particularly the mothers, do not love and care for their children these last are not likely to live, with the result that the fine moral quality of mother-love is strengthened in the race by its very survival value. The males of a race may dominate and exploit the females, making them do all the work, but the process cannot be carried too far, at any rate with the young females, for if they are overburdened they will not be able to raise the next generation, hence the moral qualities of chivalry, politeness and consideration for women have considerable survival value.

Similarly in conducting his daily life a man can lie, cheat and swindle, which may make him prosperous; on the other hand, a reputation for virtue and honesty in dealing is more likely to do so, which is borne out by the success of the large Quaker business houses, whose fortunes were founded by honest dealing in the eighteenth century, when business morality was not high. Peace of mind and happiness are more likely to arise from fair dealing than from foul and it is at least likely that the contented in mind have more children and raise them more successfully than do the discontented; this has a biological effect and is perhaps the reason why there are more relatively honest people in the world than dishonest ones. But man, as Hamlet implies, is not a beast, and apart from any worldly success arising from the virtues, man appreciates the desirability of good behaviour for its own sake and for the satisfaction of that inner compulsion, the soul, for the majority need to use 'that capability and god-like reason' to advantage. Honest dealing had a direct survival value in the early days of Bartons End, for a man or woman could be hanged for the theft of goods valued at over five shillings; there may well have been no genetic difference involved, for we do not know if the, admittedly fewer, descendants of the hanged really would be more likely to steal than the unconvicted. The majority of the humans in the house were fundamentally kind, and indeed mutual trust between master and man helped establish the farm, hence this moral quality certainly had biological effects, the honest dealing and inward satisfaction leading to contentment and a measure of prosperity, by the standards of the age, for the farm and its dependants.

In those days the families of the well-to-do were more likely to survive in greater numbers than those of the poor as the former at least had adequate food. Today wealth usually has an adverse effect on survival, the rich having fewer children than the poor.

The secondary differences between the sexes in animals have arisen mostly through sexual selection, the males or females respectively desiring some one particular characteristic in the opposite sex more than another; for instance female crickets would respond to the loudest-singing male crickets and had no need to make any noise themselves. Differing secondary sexual characteristics, with little or no survival value, may thus develop in a species, as for instance the smooth cheek of a woman and the hairy one of a man. The human generations during the life of Bartons End were too few for the race to be much affected genetically by these higher qualities. However the men did select their wives, and the women accept their husbands, with some appreciation of beauty and manliness in mind, for, even if Susannah Aylesford was as 'beautiful as an heiress with steel shares', Prudence Sefton and Anne Cobb were poor but typical Kentish beauties. In fact, an appreciation of beauty and historical values actually saved the house and its life from decay, for, as has been told, Robert Dunchester was so struck with these qualities in it, that he purchased it in 1909 at a time when the roof had fallen in; such values still shape much of the life there today.

Intellectual ability and mechanical skill have immense survival value for the race, enabling man to fill empty continents with his species and feed millions where only thousands were fed before, but the balance between success and failure is precarious; it is quite possible that some small change will extinguish man at Bartons End, or even as a species on the earth itself. It is difficult to predict the state of life in the house, but its quantity, except at night, is likely to be high and to remain so during the near future. Bartons End, as an old people's home, is a success.

Two of the three Frazer children, Phillip, Caroline and Sloane Myfanwy III, have married so far but this last daughter has three children, two of whom, Phillip, aged fifteen, and Sloane Myfanwy, aged ten, already show an interest in the old house. If their uncle has no children or their aunt does not marry they are likely to inherit the property and, one hopes, preserve it.

Whether the place can continue to thrive on its present lines is debatable. The population of Britain is aging, so the demand for comfortable homes for the old is increasing. But also increasing is the cost of running such places, so that fewer people will be able to afford the expense. Whether the comparatively young in a population will continue to support the old is a moot point. For instance, many large companies now have more pensioners than employees. But it is hoped that the beauty and appeal of Bartons End, and its long history, will somehow lead to its preservation.

It could be held that there are some low fertility genes, introduced from the Frazers, in the Thompson make-up, as two of the children, Phillip and Caroline, now 47 and 45 years old respectively, have not had children. Caroline has not married, but this is not an impossible barrier to child-bearing. Low fertility is a genetic factor whether it be due to some physical defect or an actual desire to avoid giving birth. The genes carrying such characteristics may well have arisen in the course of ages due to the survival value of small families, whether the creature be insect or mammal. Human fertility is very variable and the low-fertility genes could be associated with a high survival value. The very fact that they persist in the race suggests this. For instance, Louise Dunchester might well have had a child by means of medical intervention, which would result in her low-fertility genes being perpetuated in her child. Should a large number of the sub-fecund in the population have children by this means the incidence of low-fertility genes in the country's inhabitants would increase. People as a whole would become less fertile and in time the situation could reach the critical point, where there were not enough meetings between fertile males and females for the race to be continued.

One imagines that man would again intervene before this situation arose and save himself once more by the skin of his teeth for he has survived many catastrophes—flood, ice-ages, pestilence, massacre, war and famine—so he should be able to survive this too. This intervention might merely be the re-colonizing of Kent by a more fertile race of men from overseas, for all the time there are men on earth none of the space habitable by them is likely to be surrendered to any other creature.

Man, however, does now possess the power of destroying himself entirely by means of a major nuclear war. Whilst he and many of the higher animals could be wiped out by such a disaster, life itself is unlikely to be entirely destroyed and as the killing radiation died away, the evolutionary process would take up its task again, from whatever remnants of life were left and once more the site at least at Bartons End would fill with life, but this does not necessarily mean that man would arise again—Dr G. G. Simpson thinks this unlikely, for man's ancient ancestral forms have vanished, together with the sequence of biological and physical conditions that gave this pattern to our evolution. Man obtained his dominant position because of his intelligence, the adaptation of his forefeet to holding things and to a close co-operation of hand and eye. Many other mammals outside the apes and monkeys have these last two qualities, for instance squirrels, rats and mice—even some birds can use a foot as a hand, so that a life of this nature might well fill the space should man vacate it.

There is very little space that life could not fill and the only future development, unlikely though not impossible, that Simpson foresees is the colonization of the air—'. . . a truly aerial flora and fauna of organisms living and reproducing in air as a medium, as seaweeds and fishes do in water. . . .' Charming as it sounds, it might turn out to be most inconvenient to us.

There has been much 'throwing about of brains' in this book. The house stands in its quiet garden, a little apart from the farm, unostentatious and a monument to man's efforts. Its present life is not so varied as it was, but the supplies it draws in now affect the lives of other men and animals in far wider fields than when it was new. The house does not only receive but sends out as well, for using man's unique gift—the appreciation of past, present and future—some of the patients at Bartons End, particularly Dr Dhusti, export their ideas in books, articles and broadcasts to stimulate their fellows in fields as wide as those from which they draw their food. 'What a piece of work is man! How noble in reason! How infinite in faculty! in form, in moving, how express and admirable! in action how like an angel! in apprehension how like a god! the beauty of the world! the paragon of animals. . . .' This, written forty-six years after Bartons End was built, had an ironic tone then, is not quite true today, but can become true if man will only use his gifts with sense and goodwill.

Appendix

The Names of the Animals

One of the great contributions to human progress from the earliest times was the naming of things. As language developed, men communicated their thoughts to each other and were thus able to co-operate more effectively. It is a big step forward to pass from the general cry of alarm, meaning 'danger', found among so many animals, to that of 'lion' or 'mammoth', specifying the particular kind of danger, and it enabled the human animal to dominate in the world. As civilization developed it became more and more important to know just exactly what animal, or other form of life, was meant by the common words such as 'mouse', 'bean' or 'beetle'. Clearly there are dozens of kinds of mice, hundreds of kinds of beans and thousands of beetles.

During the sixteenth century an increasing awareness of life was starting among philosophers and much confusion was caused by the vagueness in references to animals. This led to the invention of descriptive names, usually in Latin, but it was not until the mid-eighteenth century that the binomial system was developed by the great Linnaeus. One name was for the genus to which the animal belonged and the other was specific to that particular animal only. *Culex pipiens* was the name Linnaeus gave to a particular gnat, and it can mean no other. The names are usually descriptive of some feature of the animal: *Culex* is the Latin for gnat and was applied to a particular genus of gnats only and *pipiens*, meaning whispering, is a particular gnat in that genus. In an age when most educated people knew Latin and Greek the meanings of the scientific names were more commonly known than today, when many, in fact most, biologists, do not know what these words mean, that slip so glibly and continuously from their tongues. Below is a list of the animals mentioned in this book, with their scientific names and a meaning that can be given to these words.

LIST OF ANIMALS

English Name	Scientific Name	Meaning
	CHAPTER THREE	
Lyctus or Powder Post beetles	*Lyctus linearis L. brunneus*	Linear Lyctus. Dark brown Lyctus (a town in Crete)
Furniture beetle	*Anobium punctatum*	Pitted dead-shammer
	CHAPTER FOUR	
Man	*Homo sapiens*	Wise man
	CHAPTER FIVE	
Death Watch beetle	*Xestobium rufovillosum*	Red-haired polished life
Black Clerid beetle	*Corynetes caeruleus*	Dark-blue mace bearer
Soft brown beetle	*Opilo mollis*	Soft shepherd
Theocolax	*Theocolax formiciformis*	Ant-like God-flatterer
Spathius	*Spathius pedestris*	The walking bladed or spatula-like creature
	CHAPTER SIX	
Mouse	*Mus musculus domesticus*	Domestic little-mouse mouse
Dog	*Canis familiaris*	Domestic dog
Cat	*Felis maniculata*	Small-handed cat
Noctule bat	*Nyctalus noctula*	Drowsy night creature
Pipistrelle bat	*Pipistrellus pipistrellus*	Little-bat little bat
	CHAPTER SEVEN	
Cinnabar moth	*Hypocrita jacobae*	Ragwort actor
The hanger spider	*Pholcus phalangioides*	Jointed squint-eyed creature
The long-legged spider	*Tegenaria domestica*	Domestic mat creature
The tiny pink spider	*Oonops pulcher*	Beautiful egg-face
The fierce Ciniflo	*Ciniflo ferox*	Fierce hair-curler
The mouse-grey	*Herpyllus blackwalli*	Blackwall's creeper
The long-legged spider	*Tegenaria parietina*	Mural tegeate
The short-legged brown	*Steatoda bipunctata*	Two-spotted fat one

CHAPTER EIGHT

House sparrow	*Passer domesticus*	Domestic sparrow
House martin	*Delichon urbica*	City swallow (*Delichon* is an anagram of *Chelidon*)
Cuckoo	*Cuculus canorus*	Tuneful cuckoo
Goshawk	*Accipiter gentilis*	People's hawk
Sparrow hawk	*Accipiter nisus*	Nisus's hawk
Peregrine	*Falco peregrinus*	Foreign falcon

CHAPTER NINE

Common clothes moth	*Tineola biselliella*	Double-seated little moth
White-shouldered house moth	*Endrosis sarcitrella*	In the dewy flesh
Bacon beetle	*Dermestes ladarius*	Bacon or leather eater
Leather beetle	*D. maculatus*	Spotted leather eater
Common carpet beetle	*Anthrenus scrophulariae*	Scrofulous without complaint
Spider beetle	*Ptinus fur*	Thieving flier
Grain weevil	*Calandra granaria*	Grain weevil
Cellar beetle or meal worm	*Tenebrio molitor*	Striving lover of darkness
Dark meal worm	*T. obscurus*	Dark lover of darkness
Flour beetle	*Tribolium confusum*	Disorderly three-pointed creature
Indian meal moth	*Plodia interpunctella*	Little pitted Plodia
Mediterranean flour moth	*Ephestia kuhniella*	Kuhn's fireside creature

CHAPTER TEN

Flour mite	*Tyroglyphus farinae*	Flour cheese borer
Predacious mite	*Cheyletus eruditus*	Learned Cheylean
False scorpions	*Chelifer cancroides* C. museorum	Crab-like claw-bearer Museum claw-bearer
Cheese mite	*Tyrolichus casei*	Cheese cheese-licker
Furniture mite	*Glycyphagus domesticus*	Domestic sugar-eater
Silverfish	*Lepisma saccharina*	Sugar scale
Firebrat	*Thermobia domestica*	Domestic hot-liver

Booklouse	*Trogium pulsatorium* *Liposcelis granicola*	Knocking gnawer Grain-growing fat-legged creature
House-cricket	*Gryllus domesticus*	Domestic grasshopper
Oriental cockroach or black beetle	*Blatta orientalis*	Eastern cockroach
German cockroach or steam-fly	*Blatella germanica*	German little cockroach
Vinegar eelworm	*Anguillula aceti*	Vinegar eel

CHAPTER ELEVEN

Cabbage white	*Pieris brassicae*	Cabbage Pieris
Apanteles parasite	*Apanteles glomeratus*	All spinning a roll
Human flea	*Pulex irritans*	Irritant flea
Cat flea	*Ctenocephalides felis*	Cat comb-headed creature
Dog flea	*C. canis*	Dog do.
Rat flea	*Xenopsylla cheopis*	Cheops stranger-flea
Mouse flea	*Leptopsylla segnis*	Lazy small flea
Rabbit flea	*Spilopsyllus cuniculi*	Rabbit spotted flea
Human tapeworm, beef tapeworm	*Taenia saginata*	Fattened tapeworm
Pork tapeworm	*Taenia solium*	Usual tapeworm
Dog tapeworm	*Echinococcus granulosus*	Small grained spiked creature
Cat liver-fluke	*Opisthorcis tenuicollis*	Thin-limbed testicles at the back creature
Body-louse	*Pediculus humanus corporis*	Human body louse
Head-louse	*P. humanus capitis*	Human head louse
Pubic-louse or crab	*Phthirus pubis*	Pubic louse
Dog louse, sucking	*Linognathus setosus*	Hairy linen-jawed
Dog louse, biting	*Trichodectes canis*	Dog hair
Bird louse, on Passerinformes	*Colpocephalum, Myrsidea*	Bosom headed, Mercer
Feather-lice	*Menacanthus, Ricinus, Bruelia, Sturnidoecus, Penenirmus, Philopterus*	Moon-horn, Castor oil, Bruel's, Stealing, Dweller, Wing-lover
on Apodiformes	*Dennyus, Eureum,* species	Dennis's, Fair-flowering

Bed-bug	*Cimex lectularius*	Bed-bug
Swallow-bug	*Oeciacus hirundinis*	Swallow house-pointed creature
Assassin or fly-bug	*Reduvius personatus*	Masked hangnail
House-martin flea	*Ceratophyllus hirundinis*	Swallow's horn-leafed creature
House-sparrow flea	*C. fringillae*	Fringilla horn-leafed creature
Beetle predator	*Microglotta* species	Small-tongued
Swallow house-fly	*Stenepteryx hirundinis*	Swallow's narrow-winged creature
Black flies	*Simulium* species	Imitator
Birdbottle	*Protocalliphora azurea*	Sky-blue first beauty carrier
Dog roundworm	*Toxacta canis*	Dog arrow-poison
Dog tapeworm	*Taenia pisiformis* *Diplydium caninum*	Pea-shaped tapeworm Dog's double-fold
Dog bladder-worm	*T. echinococcus*	Spiky tapeworm
Dog long tapeworm	*Dibothriocephalus latus*	Broad double-small-hollow-headed creature
Dog mange, follicular common	*Demodex folliculorum* *Sarcoptes scabies v. canis*	Follicular body-worm Rough flesh cutter
Scabies or itch of man	*Sarcoptes scabies v. humanis*	do.

CHAPTER TWELVE

Common house-fly	*Musca domestica*	Domestic fly
Lesser house-fly	*Fannia canicularis*	Dog Fannia
Bluebottle	*Calliphora erythrocephala*	Red-headed beauty bearer
Greenbottle	*Lucilia sericata*	Silk-clad Lucilia
Cluster-fly	*Pellenia rudis*	Wild flour creature
Mosquito	*Anopheles maculipennis*	Spotted winged harmful creature
	Theobaldia annulata	Ringed Theobaldia
House-gnat	*Culex pipiens*	Whispering gnat
Black or garden ant	*Acanthomyops (Lasius) niger*	Black-thorn fly
Pharaoh's ant	*Monomorium pharaonis*	Pharaoh's single member

Argentine ant	*Iridomyrmex humilis*	Small Isis ant
Earwig	*Forficula auricularia*	Ear scissors
Fat dormouse	*Glis glis*	Dormouse dormouse
Common wasp	*Vespula vulgaris*	Common little wasp
German wasp	*V. germanica*	German little wasp
Red wasp	*V. rufa*	Red little wasp
Wood wasp	*V. sylvestris*	Wood little wasp
Norwegian wasp	*V. norvegica*	Norwegian little wasp
Hornet wasp	*Vespa crabro*	Hornet wasp
Herald moth	*Scoliopterya libatrix*	Libation pouring crooked winged creature
Silver-Y moth	*Plusia gamma*	Gamma rich creature
March moth	*Alsophila aescularia*	Oak grove lover
Small tortoiseshell	*Vanessa urticae*	Nettle Vanessa
Peacock butterfly	*V. io*	Io Vanessa
Ladybird	*Adalia bipunctata*	Double-spotted unhurt creature
Common toad	*Bufo vulgaris*	Common toad
Natterjack	*B. calamita*	Reedy toad

Bibliography

Many works have been consulted in the preparation of this book: the principal ones are given below, and the reader who requires more information on any point will find it useful to turn to some of them.

Chapters 2 and 4 Building ... the Home. Bartons End
BRAUN, H. *The Story of English Architecture*. Faber, 1950.
BRIGGS, M. S. *The English Farmhouse*. Batsford, 1953.
FUSSELL, G. E *The English Rural Labourer*. Batchworth, 1949.
PLAT, H. ED FUSSELL, G. E. *Delights for Ladies*. Crosby Lockwood, 1955.
History of Technology, Vol. III. Clarendon Press, 1958.
INNOCENT, C. F. *The Development of English Building Construction*. Cambridge University Press, 1916.
JEKYL, G. and JONES, S. R. *Old English Household Life*. Batsford, 1939.
MOXON, J. *Mechanick Exercises, or the Doctrines of Handy Works*, 1677–79.

Chapters 3 and 5 Wood-borers
ANON. *Furniture Beetles*. British Museum Economic Series No. 11.
HICKEN, N. E. *Woodworm*. Hicken, London, 1954.
IMMS, A. D. *A General Textbook of Entomology*. Methuen, 9th ed. 1957.
LEPESME, P. *Les Coléoptères des Denrées Alimentaires*. P. Chevalier, Paris, 1944.
ROEDER, K. D. Ed. *Insect Physiology*, J. Wiley, New York. Chapman and Hall, London, 1953.
WIGGLESWORTH, V. B. *The Principles of Insect Physiology*. Methuen, 5th ed. 1953.

Chapter 6 Large and Small Mammals
HINTON, M. A. C. *Rats and Mice*. British Museum Economic Series No. 8.
HUBBARD, C. L. B. *An Introduction to the Literature of British Dogs*. Hubbard, Ponterwyd, 1949.
LANCUM, F. H. *Wild Mammals and the Land*. H.M.S.O., London, 1951.
MATTHEWS, L. H. *British Mammals*. Collins, London, 1952.
SOUTHERN, H. N. *House Mice*, Vol. 3 of *Control of Rats and Mice*. Chitty and Southern. Clarendon Press, Oxford, 1954.
THEOBALD, F. V. *A Text-Book of Agricultural Zoology*. Blackwood, London, 1919.
VESEY-FITZGERALD, B. *British Bats*. Methuen, London, 1949.
—— Ed. *The Book of the Dog*. Nicholson and Watson, 1948.
—— *Cats*. Penguin Books, 1957.

Chapter 7   Spiders

BRISTOWE, W. S. *The Comity of Spiders*. Ray Society, London, 1939, 1941.
DISJONVAL, Q. *De l'Aranéologie*. Fuchs, Paris, 1797.
LOCKET, G. H. and MILLIDGE, A. F. *British Spiders*. Ray Society, London, 1951, 1953.
THEOBALD, F. V. *Op. cit.*

Chapter 8   Birds

BURTON, M. 'A Possible Explanation of the Phoenix Myth.' *New Scientist*, 27th June, 1957.
CARTHY, J. D. *Animal Navigation*. Allen and Unwin, 1956.
FLETCHER, R. *Instinct in Man*. Allen and Unwin, 1957.
LACK, D. *Life of the Robin*. H. F. and G. Witherley, 1944.
NICHOLSON, E. M. *Birds and Man*. Collins, 1951.
—— *How Birds Live*. Williams and Norgate, 2nd ed., 1937.
ROTHSCHILD, M. and CLAY, T. *Fleas, Flukes and Cuckoos*. Collins, 1952.
THEOBALD, F. V. *Op. cit.*
YOUNG, J. Z. *Vertebrates* Oxford, 1950.

Chapters 9 and 10   The Dry-litter Community. Adaptors

AUSTEN, E. E. and MCKENNY. *Clothes Moths and House Moths*. Economic Series No. 14, British Museum, 1935.
GUENAUX, G. *Entomologie et Parasitologie Agricole*. J. Balière, Paris, 1922.
MIALL, L. C. and DENNY, A. *The Cockroach*. Lovell Reeve, 1886.
ROEDER, K. D. *Op. cit.*
VOM WACHENDORF, F. *L'Homme et les Fléaux*. La Table Ronde, Paris, 1954.

Chapter 11   Third-hand

BUSVINE, J. R. *Insects and Hygiene*. Methuen, London, 1951.
GUENAUX G. *Op. cit.*
LAPAGE, G. *Animals Parasitic in Man*. Pelican (Penguin Books), 1957.
ROTHSCHILD, M. and CLAY, T. *Op. cit.*
THEOBALD, F. V. *Op. cit.*
VESEY-FITZERALD, B. *Op. cit.*

Chapter 12 and 13   Visitors. Past, Present and Future

FUSSELL, G. E. *Op. cit.*
GOODHART, C. B. 'The Future of Human Fertility,' *New Scientist*, 12th December, 1957.
JEKYL, G. and JONES, S. R. *Op. cit.*
JOHNSTONE, J. *The Essentials of Biology*. Arnold, 1932.
KROGH, A. *The Respiratory Exchange of Animals and Man*. Longmans, Green, 1916.

MCFARLAND, D. (Ed.) *The Oxford Companion to Animal Behaviour*. Oxford University Press. 1981.

PLAT, H. *Op. cit.*

SIMPSON, G. G. *The Meaning of Evolution*. Yale University Press, 1949.

TREVELYAN, G. M. *English Social History*. Longmans, Green, 1946.

Index

Numbers in italics refer to illustrations